Complex Surveys

Complex Surveys

A Guide to Analysis Using R

Thomas Lumley

University of Washington
Department of Biostatistics
Seattle, Washington

A John Wiley & Sons, Inc., Publication

Published by John Wiley & Sons, Inc., Hoboken, New Jersey.
Published simultaneously in Canada.

Library of Congress Cataloging-in-Publication Data:

Lumley, Thomas, 1969–
  Complex surveys : a guide to analysis using R / Thomas Lumley.
     p. cm.
  Includes bibliographical references and index.
  ISBN 978-0-470-28430-8 (pbk.)
  1. Mathematical statistics—Data processing. 2. R (Computer program language) I. Title.
  QA276.45.R3L86 2010
  515.0285—dc22   2009033999

Printed in the United States of America.

10 9 8 7 6 5 4 3 2

CONTENTS

v

Acknowledgments

Most of this book was written while I was on sabbatical at the University of Auckland and the University of Leiden. The Statistics department in Auckland and the Department of Clinical Epidemiology at Leiden University Medical Center were very hospitable and provided many interesting and productive distractions from writing.

I had useful discussions on a number of points with Alastair Scott and Chris Wild. Bruce Psaty, Stas Kolenikov, and Barbara McKnight gave detailed and helpful comments on a draft of the text. The "'s" interpretation of the $ operator came from Ken Rice. Hadley Wickham explained how to combine city and state data in a single map. Paul Murrell made some suggestions about types of graphics to include. The taxonomy of regression predictor variables is from Scott Emerson. I learned about some of the references on reification from Cosma Shalizi's web page. The students and instructors in STAT/CSSS 529 (Seattle) and STATS 740 (Auckland) tried out a draft of the book and pointed out a few problems that I hope have been corrected.

Some financial support for my visit to Auckland was provided by Alastair Scott, Chris Wild, and Alan Lee from a grant from the Marsden Fund, and my visit to Leiden was supported in part by Fondation Leducq through their funding of the LINAT collaboration. My sabbatical was also supported by the University of Washington.

The survey package has benefited greatly from comments, questions, and bug reports from its users, an attempt at a list is in the THANKS file in the package.

Preface

This book presents a practical guide to analyzing complex surveys using R, with occasional digressions into related areas of statistics. Complex survey analysis differs from most of statistics philosophically and in the substantive problems it faces. In the past this led to a requirement for specialized software and the spread of specialized jargon, and survey analysis became separated from the rest of statistics in many ways. In recent years there has been a convergence of ways. All major statistical packages now include at least some survey analysis features, and some of the mathematical techniques of survey analysis have been incorporated in widely-used statistical methods for missing data and for causal inference.

More importantly for this book, researchers in the social science and health sciences are increasingly interested in using data from complex surveys to conduct the same sorts of analyses that they traditionally conduct with more straightforward data. Medical researchers are also increasingly aware of the advantages of well-designed subsamples when measuring novel, expensive variables on an existing cohort.

This book is designed for readers who have some experience with applied statistics, especially in the social sciences or health sciences, and are interested in learning about survey analysis. As a result, we will spend more time on graphics, regression modelling, and two-phase designs than is typical for a survey analysis text. I have presented most of the material in this book in a one-quarter course for graduate students who are not specialist statisticians but have had a graduate-level introductory course in applied statistics, including linear and logistic regression. Chapters 1–6 should be of general interest to anyone wishing to analyze complex surveys. Chapters 7–10 are, on average, more technical and more specialized than the earlier material, and some of the content, particularly in Chapter 8, reflects recent research.

The widespread availability of software for analyzing complex surveys means that it is no longer as important for most researchers to learn a list of computationally convenient special cases of formulas for means and standard errors. Formulas will be presented in the text only when I feel they are useful for understanding concepts; the appendices present some additional mathematical and computational descriptions that will help in comparing results from different software systems. An excellent reference for statisticians who want more detail is *Model Assisted Survey Sampling* by Särndal, Swensson, and Wretman [151]. Some of the exercises presented at the end of each chapter require more mathematical or programming background, these are indicated with a ★. They are not necessarily more difficult than the unstarred exercises.

This book is designed around a particular software system: the survey package for the R statistical environment, and one of its main goals is to document and explain this system. All the examples, tables, and graphs in the book are produced with R, and code and data for you to reproduce nearly all of them is available. There are three reasons for choosing to emphasize R in this way: it is open-source software, which makes it easily available; it is very widely known and used by academic statisticians, making it convenient for teaching; and because I designed the survey package it emphasizes the areas I think are most important and readily automated about design-based inference. For other software for analyzing complex surveys, see the comprehensive list maintained by Alan Zaslavsky at http://www.hcp.med. harvard.edu/statistics/survey-soft/.

There are important statistical issues in the design and analysis of complex surveys outside design-based inference that I give little or no attention to. Small area estimation and item response theory are based on very different areas of statistics, and I think are best addressed under spatial statistics and multivariate statistics, respectively. Statistics has relatively little positive to say about about non-sampling error, although I do discuss raking, calibration, and the analysis of multiply-imputed data. There are also interesting but specialized areas of complex sampling that are not covered in the book (or the software), mostly because I lack experience with their application. These include adaptive sampling techniques, and methods from ecology such as line and quadrat sampling.

Code for reproducing the examples in this book (when not in the book itself), errata, and other information, can be found from the web site: http://faculty. washington.edu/tlumley/svybook. If you find mistakes or infelicities in the book or the package I would welcome an email: tlumley@u.washington.edu.

Acronyms

N	Population size
N_k	Population size for stratum k
n	Sample size
n_k	Sample size for stratum k
π_i	Sampling probability for unit i
$\check{X}_i$	Weighted observation X_i / π_i
$\hat{T}$	Estimate of T
IPW	Inverse-Probability Weighted
IPTW	Inverse-Probability of Treatment Weighted
fpc	Finite-population correction (to standard errors)
$E[\,]$	Expected value
$\Pr[\,]$	Probability
$\mathbb{I}$	An influence function

CHAPTER 1

BASIC TOOLS

In which we meet the probability sample and the R language.

1.1 GOALS OF INFERENCE

1.1.1 Population or process?

The mathematical development for most of statistics is *model-based*, and relies on specifying a probability model for the random process that generates the data. This can be a simple parametric model, such as a Normal distribution, or a complicated model incorporating many variables and allowing for dependence between observations. To the extent that the model represents the process that generated the data, it is possible to draw conclusions that can be generalized to other situations where the same process operates. As the model can only ever be an approximation, it is important (but often difficult) to know what sort of departures from the model will invalidate the analysis.

Complex Surveys: A Guide to Analysis Using R. By Thomas Lumley
Copyright © 2010 John Wiley & Sons, Inc.

The analysis of complex survey samples, in contrast, is usually *design-based*. The researcher specifies a population, whose data values are unknown but are regarded as fixed, not random. The observed sample is random because it depends on the random selection of individuals from this fixed population. The random selection procedure of individuals (the *sample design*) is under the control of the researcher, so all the probabilities involved can, in principle, be known precisely. The goal of the analysis is to estimate features of the fixed population, and design-based inference does not support generalizing the findings to other populations.

In some situations there is a clear distinction between population and process inference. The Bureau of Labor Statistics can analyze data from a sample of the US population to find out the distribution of income in men and women in the US. The use of statistical estimation here is precisely to generalize from a sample to the population from which it was taken.

The University of Washington can analyze data on its faculty salaries to provide evidence in a court case alleging gender discrimination. As the university's data are complete there is no uncertainty about the distribution of salaries in men and women in this population. Statistical modelling is needed to decide whether the differences in salaries can be attributed to valid causes, in particular to differences in seniority, to changes over time in state funding, and to area of study. These are questions about the process that led to the salaries being the way they are.

In more complex analyses there can be something of a compromise between these goals of inference. A regression model fitted to blood pressure data measured on a sample from the US population will provide design-based conclusions about associations in the US population. Sometimes these design-based conclusions are exactly what is required, e.g., there is more hypertension in blacks than in whites. Often the goal is to find out why some people have high blood pressure: is the racial difference due to diet, or stress, or access to medical care, or might there be a genetic component?

1.1.2   Probability samples

The fundamental statistical concept in design-based inference is the *probability sample* or *random sample*. In everyday speech, "taking a random sample" of 1000 individuals means a sampling procedure when any subset of 1000 people from the population is equally likely to be selected. The technical term for this is a "simple random sample". The Law of Large Numbers implies that the sample of 1000 people is likely to be representative of the population, according to essentially any criteria we are interested in. If we compute the mean age, or the median income, or the proportion of registered Republican voters in the sample, the answer is likely to be close to the value for the population.

We could also end up with a sample of 1000 individuals from the US population, for example, by taking a simple random sample of 20 people from each state. On many criteria this sample is unlikely to be representative, because people from states with low populations are more likely to be sampled. Residents of these states have a similar age distribution to the country as a whole but tend to have lower incomes and be more politically conservative. As a result the mean age of the sample will be close to the mean age for the US population, but the median income is likely to be lower, and the proportion of registered Republican voters higher than for the US population. As long as we know the population of each state, this *stratified random sample* is still a probability sample. Yet another approach would be to choose a simple random sample of 50 counties from the US and then sample 20 people from each county. This sample would over-represent counties with low populations, which tend to be in rural areas. Even so, if we know all the counties in the US, and if we can find the number of households in the counties we choose, this is also a probability sample.

It is important to remember that what makes a *probability sample* is the procedure for taking samples from a population, not just the data we happen to end up with.

The properties we need of a sampling method for design-based inference are as follows:

1. Every individual in the population must have a non-zero probability of ending up in the sample (written π_i for individual i)

2. The probability π_i must be known for every individual who does end up in the sample.

3. Every pair of individuals in the sample must have a non-zero probability of both ending up in the sample (written π_{ij} for the pair of individuals (i, j)).

4. The probability π_{ij} must be known for every pair that does end up in the sample.

The first two properties are necessary in order to get valid population estimates; the last two are necessary to work out the accuracy of the estimates. If individuals were sampled independently of each other the first two properties would guarantee the last two, since then $\pi_{ij} = \pi_i \pi_j$, but a design that sampled one random person from each US county would have $\pi_i > 0$ for everyone in the US and $\pi_{ij} = 0$ for two people in the same county. In the survey package, as in most software for analysis of complex samples, the computer will work out π_{ij} from the design description, they do not need to be specified explicitly.

The world is imperfect in many ways, and the necessary properties are present only as approximations in real surveys. A list of residences for sampling will include some that are not inhabited and miss some that have been newly constructed. Some people (me, for example) do not have a landline telephone, others may not be at home or may refuse to answer some or all of the questions. We will initially ignore these problems, but aspects of them are addressed in Chapters 7 and 9.

1.1.3 Sampling weights

If we take a simple random sample of 3500 people from California (with total population 35 million) then any person in California has a 1/10000 chance of being sampled, so $\pi_i = 3500/3500000 = 1/10000$ for every i. Each of the people we sample represents 10000 Californians. If it turns out that 400 of our sample have high blood pressure and 100 are unemployed, we would expect $400 \times 10000 = 4$ million people with high blood pressure and $100 \times 10000 = 1$ million unemployed in the whole state. If we sample 3500 people from Connecticut (population $3,500,000$), all the sampling probabilities are equal to $3500/3500000 = 1/1000$, so each person in the sample represents 1000 people in the population. If 400 of the sample had high blood pressure we would expect $400 \times 1000 = 400000$ people with high blood pressure in the state population.

The fundamental statistical idea behind all of design-based inference is that an individual sampled with a sampling probability of π_i represents $1/\pi_i$ individuals in the population. The value $1/\pi_i$ is called the *sampling weight*.

This weighting or "grossing up" operation is easy to grasp for a simple random sample where the probabilities are the same for every one. It is less obvious that the same rule applies when the sampling probabilities can be different. In particular, it may not be intuitive that the sampling probabilities for individuals who were not sampled do not need to be known.

Consider measuring income on a sample of one individual from a population of N, where π_i might be different for each individual. The estimate ($\hat{T}_{\text{income}}$) of the total income of the population (T_{income}) would be the income for that individual multiplied by the sampling weight:

$$\hat{T}_{\text{income}} = \frac{1}{\pi_i} \times \text{income}_i.$$

This will not be a very good estimate, since it is based on only one person, but it will be *unbiased*: the expected value of the estimate will equal the true population total. The expected value of the estimate is the value of the estimate when we select person i, times the probability of selecting person i, added up over all people in the population

$$
\begin{aligned}
E\left[\hat{T}_{\text{income}}\right] &= \sum_{i=1}^{N} \frac{1}{\pi_i} \times \text{income}_i \times \pi_i \\
&= \sum_{i=1}^{N} \text{income}_i \\
&= T_{\text{income}}.
\end{aligned}
$$

The same algebra applies with only slightly more work to samples of any size. The $1/\pi_i$ sampling weights used to construct the estimate cancel out the π_i probability that this particular individual is sampled. The estimator of the population total is called the Horvitz–Thompson estimator [63] after the authors who proposed the most general form and a standard error estimate for it, but the principle is much older.

Estimates for any other population quantity are derived in various ways from estimates for a population total, so the Horvitz–Thompson estimator of the population total is the foundation for all the analyses described in the rest of the book. Because of the importance of sampling weights and the inconvenience of writing fractions it is useful to have a notation for the weighted observations. If X_i is a measurement of variable X on person i, we write

$$\check{X}_i = \frac{1}{\pi_i} X_i.$$

Given a sample of size n the Horvitz–Thompson estimator $\hat{T}_X$ for the population total T_X of X is

$$\hat{T}_X = \sum_{i=1}^{n} \frac{1}{\pi_i} X_i = \sum_{i=1}^{n} \check{X}_i. \tag{1.1}$$

The variance estimate is

$$\widehat{\text{var}}\left[\hat{T}_X\right] = \sum_{i,j} \left(\frac{X_i X_j}{\pi_{ij}} - \frac{X_i}{\pi_i} \frac{X_j}{\pi_j} \right). \tag{1.2}$$

Knowing the formula for the variance estimator is less important to the applied user, but it is useful to note two things. The first is that the formula applies to any design, however complicated, where π_i and π_{ij} are known for the sampled observations. The second is that the formula depends on the pairwise sampling probabilities π_{ij}, not just on the sampling weights; this is how correlations in the sampling design enter the computations. Some other ways of writing the variance estimator are explored in the exercises at the end of this chapter.

Other meanings of "weights"   Statisticians and statistical software use the term 'weight' to mean at least three different things.

sampling weights A sampling weight of 1000 means that the observation represents 1000 individuals in the population.

precision weights A precision (or inverse-variance) weight of 1000 means that the observation has 1000 times lower variance than an observation with a weight of 1.

frequency weights A frequency weight of 1000 means that the sample contains 1000 identical observations and space is being saved by using only one record in the data set to represent them.

In this book, weights are always sampling weights, $1/\pi_i$. Most statistical software that is not specifically designed for survey analysis will assume that weights are precision weights or frequency weights. Giving sampling weights to software that is expecting precision weights or frequency weights will often (but not always) give correct point estimates, but will usually give seriously incorrect standard errors, confidence intervals, and p-values.

1.1.4 Design effects

A complex survey will not have the same standard errors for estimates as a simple random sample of the same size, but many sample size calculations are only conveniently available for simple random samples. The *design effect* was defined by Kish (1965) as the ratio of a variance of an estimate in a complex sample to the variance of the same estimate in a simple random sample [75].

If the necessary sample size for a given level of precision is known for a simple random sample, the sample size for a complex design can be obtained by multiplying by the design effect. While the design effect will not be known in advance, some useful guidance can be obtained by looking at design effects reported for other similar surveys.

Design effects for large studies are usually greater than 1.0, implying that larger sample sizes are needed for complex designs than for a simple random sample. For example, the California Health Interview Survey reports typical design effects in the range 1.4–2.0. It may be surprising that complex designs are used if they require both larger samples sizes and special statistical methods, but as Chapter 3 discusses, the increased sample size can often still result in a lower cost.

The other ratio of variances that is of interest is the ratio of the variance of a correct estimate to the incorrect variance that would be obtained by pretending that the data are a simple random sample. This ratio allows the results of an analysis to be (approximately) corrected if software is not available to account for the complex design. This second ratio is sometimes called the design effect and sometimes the misspecification effect.

That is, the design effect compares the variance from correct estimates in two different designs, while the misspecification effect compares correct and incorrect analyses of the same design. Although these two ratios of variances are not the same, they are often similar for practical designs. The misspecification effect is of relatively little interest now that software for complex designs is widely available, and it will not appear further in this book.

1.2 AN INTRODUCTION TO THE DATA

Most of the examples used in this book will be based either on real surveys or on simulated surveys drawn from real populations. Some of the data sets will be quite large by textbook standards, but the computer used to write this book is a laptop dating from 2006, so it seems safe to assume that most readers will have access to at least this level of computer power. Links to the source and documentation for all these data sets can be found on the web site for the book.

Nearly all the data are available to you in electronic form to reproduce these analyses, but some effort may be required to get them. Surveys in the United States tend to provide (non-identifying, anonymized) data for download by anyone, and the datasets from these surveys used in this book are available on the book's web site in directly usable formats. Access to survey data from Britain tends to require much filling in of forms, so the book's web site provides instructions on where to find the data and how to convert it to usable form. These national differences partly reflect the differences in copyright policy in the two countries. In the US, the federal government places materials created at public expense in the public domain; in Britain, the copyright is retained by the government.

You may be unfamiliar with some of the terminology in the descriptions of data sets, which will be described in subsequent chapters.

1.2.1 Real surveys

NHANES.   The National Health and Nutrition Examination Surveys have been conducted by the US National Center for Health Statistics (NCHS) since 1970. They are designed to provide nationwide data on health and disease, and on dietary and clinical risk factors. Each four-year cycle of NHANES recruits about 28000 people in a multistage sample. These participants receive an interview and a clinical exam, and have blood samples taken. Several hundred data variables are available in the public use data sets.

FRS.   The Family Resources Survey collects information on the incomes and circumstances of private households in the United Kingdom. It was designed to collect information needed by the Department for Work and Pensions. The survey first samples 1848 postcode sectors from Great Britain, stratified by geographic region and by some employment and income variables. The postcode sectors are sampled with probability proportional to the number of mailing addresses with fewer than 50 mail items per day, an estimate of the number of households. Within each postcode sector a simple random sample of households is taken. A few variables from the Scottish subset of FRS have been made available by the PEAS project at Napier University (after some modification to protect anonymity).

NHIS.   The National Health Interview Survey, conducted by the National Center for Health Statistics is the oldest of the major health-related surveys in the United States. The National Health Survey Act (1956) provided "for a continuing survey and special studies to secure accurate and current statistical information on the amount, distribution, and effects of illness and disability in the United States and the services rendered for or because of such conditions." NHIS plans to sample about 35000 households, containing about 87500 people, each year, but the survey is designed so that the results will still be useful if the sampling has to be curtailed because of budget shortfalls, as happened in 2006 and 2007. NHIS, unlike NHANES, is restricted to self-reported information and does not make clinical or biological measurements on participants. NHIS was the first major survey to include instructions for analysis using R.

SIPP.   The Survey of Income and Program Participation is a series of panel surveys conducted by the US Census Bureau, with panels of US households recruited in a multistage sampling design. The sample size has varied from about 14000 to about 37000 households. SIPP asks questions about income and about participation in government support programs such as food stamps. The same households are repeatedly surveyed over time to allow economic changes to be measured more accurately.

CHIS. The California Health Interview Survey samples households from California by random-digit dialing within geographic regions. The survey is conducted every two years and samples 40000–50000 households. Unlike the surveys above, which are conducted by government agencies, CHIS is conducted by the Center for Health Policy Research at the University of California, Los Angeles. CHIS asks questions about health, risk factors for disease, health insurance, and access to health care.

SHS. The Scottish Household Survey interviews about 31000 households every two years. Individual households are sampled in densely populated areas of Scotland; in the rest of the country a two-stage sample is used. The first stage samples census enumeration districts, which contain an average of 150 households, then the second stage samples households within these districts. The survey covers a wide range of topics such as housing, income, transport, and social services. Data from a subset of variables has been made generally available (after some further modification to protect anonymity) by the PEAS project at Napier University.

BRFSS. The Behavioral Risk Factor Surveillance System is a telephone survey of behavioral risk factors for disease. The survey is conducted by most US states using materials supplied by the National Center for Health Statistics. The number of states involved has increased from 15 in 1984 to all 50 in 2007 (plus the District of Columbia, Guam, Puerto Rico, and the US Virgin Islands) and the sample size from 12000 to 430000. It is now the world's largest telephone survey.

1.2.2 Populations

Evaluating and comparing analysis methods requires realistic data where the true answer is known. We will use some complete population data to create artificial probability samples, and compare the results of our analyses to the population values. Population data are also useful for illustrating design and preprocessing calculations that are done before the survey data reach the public use files.

Election data. Voting data for the US presidential elections is available for each county. We will try to predict the result from samples of the data and use the voting data from previous elections to improve predictions.

NWTS. Wilms' tumor is a rare childhood cancer of the kidney, curable in about 90% of cases. Most children in the United States with Wilms' tumor participate in randomized clinical trials conducted by the National Wilms' Tumor Study Group. Data from these studies [54, 38] has been used extensively in research on two-phase epidemiological studies by Norman Breslow and co-workers, and some of this data is now publically available. In our analyses the focus is on estimating the risk of relapse after initially successful treatment.

Crime in Washington.   The Washington Association of Sheriffs and Police Chiefs collects data on crimes reported to police in Washington (the state, not the city). The data are reported broken down by police district and by type of crime.

API.   The California Academic Performance Index is computed from standardized tests administered to students in California schools. In addition to academic performance data for the schools there are a wide range of socio-economic variables available. These data have been used extensively to illustrate the use of survey software by Academic Computing Services at the University of California, Los Angeles.

PBC.   Primary biliary cirrhosis is a very rare liver disease that is treatable only by transplantation. Before transplantation was available, the Mayo Clinic conducted a randomized trial of what turned out to be a completely ineffective treatment. The data from 312 participants in the trial and 106 patients who did not participate was used to create a model for predicting survival that is still used in scheduling liver transplants. As the Mayo Clinic was a major center for treatment of primary biliary cirrhosis these 418 patients represent essentially the entire population in the nearby states. The de-identified public version of the dataset was created by Terry Therneau in conjunction with his development of software for survival analysis. It has become a standard teaching and research example.

1.3   OBTAINING THE SOFTWARE

R is probably the most widely used software for statistical research and for distributing new statistical methods. The design of R is based closely on Bell Labs' S, one of the first systems for interactive statistical computing. John Chambers, the main designer of S, received the Software Systems Award from the Association for Computing Machinery

> For the S system, which has forever altered how people analyze, visualize, and manipulate data.

The drawback is, of course, that users of S and R have to alter how they analyze, visualize, and manipulate data; the learning curve may sometimes be steep. R does not have a point-and-click GUI interface, and the programming is more flexible but also more complex than the macro languages of most statistical packages.

Although all the code needed to do analyses will be presented in this book, it is not all explained in detail and readers who are not familiar with R would benefit from reading an introductory book on the language. A comprehensive list of books on R is given on the R Project web page. Fox [47] is written for social scientists and Dalgaard [37] for health scientists. Chambers [32] covers more advanced programming and design philosophy for R code.

1.3.1   Obtaining R

Windows or Macintosh users can download R from the Comprehensive R Archive Network (CRAN) at the central site, http://cran.r-project.org, or at one of many mirror sites around the world (http://cran.r-project.org/mirrors. html. Most Linux distributions provide precompiled versions of R through their package systems, and users on other Unix and Unix-like systems can easily compile R from the source code available from CRAN. New versions of R come out frequently, and you should update your installation at least once a year.

System adminstrators installing R for multiple users, or people wishing to compile R from the source code, should read the *R Installation and Administration* manual available on CRAN.

1.3.2   Obtaining the survey package

An important feature of R is the huge collection of add-on packages written by users, with the number of available packages doubling about every 18 months. In particular, R itself has no features for design-based inference and survey analysis; all the analysis features in this book come from the survey package (Lumley [99, 101]).

These packages can most easily be installed from inside R, using the Packages menu on the Windows version of R, or the Packages & Data menu on the Macintosh version. Some chapters in this book also make use of other add-on packages for graphics, imputation, and database access. These will be installed in the same way. When you use a contributed R package for published research, please cite the package (as journal policies permit). The citation() function shows the preferred citation for a package or generates a default one if the author has not specified.

The examples in this book used version 3.10-1 of the survey package and were run in R version 2.7.2. The home page for the survey package (http://faculty. washington.edu/tlumley/survey) will have information about any changes for newer versions of the package as they are released. Nearly all code should continue to run without modification, but there are likely to be small changes in the formatting of output.

1.4   USING R

This section provides a brief overview of getting data into R and doing some simple computations. Further introductory material on R can be found in Appendix B.

1.4.1   Reading plain text data

The simplest format for plain text data has one record per line with variable names in the first line of the file, with variables separated by commas. Files with this structure often have names ending .csv. Most statistical packages and databases can easily export data as comma-separated text.

```
> nwts <- read.csv("C:/svybook/nwts/nwts-share.csv")
> summary(nwts)
      trel                tsur               relaps
 Min.   : 0.01095   Min.   : 0.01095   Min.   :0.0000
 1st Qu.: 4.94182   1st Qu.: 6.24093   1st Qu.:0.0000
 Median : 9.77139   Median :10.36003   Median :0.0000
 Mean   : 9.64874   Mean   :10.32634   Mean   :0.1709
 3rd Qu.:14.01095   3rd Qu.:14.42847   3rd Qu.:0.0000
 Max.   :22.50240   Max.   :22.50240   Max.   :1.0000
[... output truncated ...]
> head(nwts)
        trel        tsur relaps dead study stage unfav.pat unfav0
1 21.88090 21.88090      0    0     3     1         1      1
2 11.28268 11.28268      0    0     3     2         0      0
3 22.11362 22.11362      0    0     3     1         1      1
4  8.02464  8.02464      0    0     3     2         0      0
5 20.49829 20.49829      0    0     3     2         0      0
6 14.39562 14.39562      1    1     3     2         0      1
[... output truncated...]
>  names(nwts)
 [1] "trel"      "tsur"      "relaps"    "dead"     "study"
 [6] "stage"     "unfav.pat" "unfav0"    "age"      "yr.regis"
[11] "specwgt"   "tumdiam"
> nrow(nwts)
[1] 3915
> ncol(nwts)
[1] 12
```

Figure 1.1   Reading in a comma-separated text file

The National Wilms' Tumor Study data are in this format. The files can be read in with the function `read.csv()`. Unlike many statistical packages, R can work with multiple data sets at the same time. This means that when a data set is read in it must be given a name so that it can be identified in the future. Naming a data set is done with the operator `<-`.

It is a good idea to check that the data have been read in correctly. One check is to compute summaries of all the variables in the data set with the `summary` function, although this is not such a good idea for survey data sets with hundreds of variables. Another check is to list the first few lines of the data set with the `head()` function. Code and R output from reading the data and performing these two checks are shown in Figure 1.1. If the file is not actually in the correct format the number of variables or their names are likely to be obviously wrong. Other simple checks are to find out the number of rows and number of columns of the data set, also shown in Figure 1.1.

The > notation at the beginning of each line is the R prompt, not part of the code to be entered. If this prompt changes to a + sign, it means that R is waiting for the line of input to be finished, which may indicate that parentheses or quotation marks have been left open on the previous line. The "Escape" key will cancel the incomplete line of input. In the examples in this book the prompt will only be shown in transcripts that include R output; examples of R code without output will omit the prompt.

1.4.2 Reading data from other packages

R can read data saved in binary formats from SPSS and Stata, and the format produced by PROC XPORT in SAS. NHANES data are now distributed in the PROC XPORT format, as are data from BRFSS. The Inter-University Consortium for Political and Social Research (ICPSR) and the SodaPop archive at Pennsylvania State University often provide data sets in Stata and SPSS formats, saving the effort needed to construct variable names and value labels for data read in as plain text.

The R functions for reading data in these formats are in the foreign package. This package is part of the R distribution, but is not automatically loaded into memory when R starts. To load the package from the package library, type

```
library(foreign)
```

When the package is loaded all its functions and help pages become available. The functions `read.xport()`, `read.dta()`, and `read.spss()` will read SAS XPORT, Stata, and SPSS files, respectively. These functions take a file name as the first argument, and `read.dta()` and `read.spss()` have other options that control the handling of dates and factors.

As an example, consider reading in the demographics file from NHANES 2003–2004, `demo_c.xpt`, which is in SAS XPORT format

```
> demo<-read.xport("~/nhanes/demo_c.xpt")
> names(demo)
 [1] "SEQN"     "SDDSRVYR" "RIDSTATR" "RIAGENDR" "RIDAGEYR"
 [6] "RIDAGEMN" "RIDAGEEX" "RIDRETH1" "RIDRETH2" "DMQMILIT"
[11] "DMDBORN"  "DMDEDUC"  "INDHHINC" "INDFMINC" "INDFMPIR"
[16] "DMDMARTL" "RIDEXPRG" "SIALANG"  "SIAPROXY" "SIAINTRP"
[21] "FIALANG"  "FIAPROXY" "FIAINTRP" "MIALANG"  "MIAPROXY"
[26] "MIAINTRP" "AIALANG"  "WTINT2YR" "WTMEC2YR" "SDMVPSU"
[31] "SDMVSTRA"
```

The ~ in the file name passed to `read.xport` means the user's home directory, so the file is in the nhanes subdirectory of the user's home directory. An example of using `read.dta()` to read Stata-format data from the California Health Interview Survey is in section 2.3.1.

R can also read data directly from relational databases, but for survey analysis it is easier to leave the data in the database as described in Appendix D.

1.4.3 Simple computations

Since more than one data set can be loaded at a time, referring to a variable requires saying which data set it is in. The demo_c.xpt data set from NHANES that was loaded above is called demo, so the age variable RIDAGEYR is called demo$RIDAGEYR. The $ is like the possessive "'s"; demo$RIDAGEYR is demo's RIDAGEYR variable.

Subsets of a variable can be indicated by

- Positive numbers: demo$RIDAGEYR[100:150] is observations 100 to 150 of the variable.

- Negative numbers: demo$RIDAGEYR[-c(1:10, 100:1000)] is all the observations except 1 to 10 and 100 to 1000. The function c() collects its arguments into a single vector.

- Logical (TRUE/FALSE) vectors: demo$RIDAGEYR[demo$RIAGENDR==1] are the ages of the men (RIAGENDR is gender). Note the use of == rather than just = for testing equality.

The repeated use of $ in the same expression can become tedious, and the example for logical subsets can be written more compactly as

```
with(demo, RIDAGEYR[RIAGENDR==1])
```

where with() specifies a particular data set as the default place to look up variables.

The $ notation allows single variables to be specified, but it is also necessary to refer to groups of variables. In the example in section 2.3.1, chis_adult[,420:499] refers to columns 420 to 499 of the California Health Interview Survey adult data set. A data set can be subscripted in both rows and columns: the numbers before the comma indicate rows and the numbers after the comma indicate columns, following the usual matrix notation in mathematics. Omitting the number before the comma means that all rows are used, and all columns when the number after the comma is omitted. Subsets of a data set can also be constructed with the subset() function, for example, kids <- subset(demo, RIDAGEYR < 18). Variables in the subset expression will first be searched for in the data set, the $ notation is not needed.

New variables can be created in a data set with the same $ notation. For example, to create a variable indicating age less than 18

```
demo$under18 <- demo$RIDAGEYR < 18
```

Missing data. Missing data are indicated by NA. It is useful to think of this as "Don't Know", so that 1+NA is NA, NA==2 is NA, and even NA==NA is NA (to test for NA use is.na()). Simple statistical functions such as mean(), sd(), and median() give NA as the result if any of their input data are missing: if you don't know the numbers, you don't know the average. These functions have an option na.rm=TRUE to ask for the missing values to be omitted.

It will often be necessary to recode values such as −9 to NA before analysis, e.g.,

```
pbc[pbc$trt == -9] <- NA
```

EXERCISES

1.1 Download an up-to-date copy of R and the survey package. Visit the book's web site to see if there are any important errata or updates.

1.2 ⋆ Work through the introductory session in the R manual *An Introduction to R.*

1.3 Each visit to the front page of a newpaper's web site has (independently) a 1/1000 chance of resulting in a questionnaire on voting intentions in a forthcoming election. Assuming that everyone who is given the questionnaire responds, why are the results not a probability sample of

a) voters?

b) readers of the newspaper?

c) readers of the newspaper's online version?

1.4 You are conducting a survey that will estimate the proportion of women who used anti-malarial insecticide-treated bed nets every night during their last pregnancy. With a simple random sample you would need to recruit 50 women in any subpopulation where you wanted a standard error of less than 5 percentage points in the estimate. You are using a sampling design that has given design effects of 2–3 for proportions in previous studies in similar areas.

a) Will you need a larger or smaller sample size than 50 for a subpopulation to get the desired precision?

b) Approximately what sample size will you need to get the desired precision?

1.5 Systematic sampling involves taking a list of the population and choosing, for example, every 100th entry in the list.

a) Which of the necessary properties of a probability sample does this procedure have?

b) For systematic sampling with a random start, the procedure would be to choose a random starting point from 1, 2, ..., 100 and then take every 100th entry starting at the random point. Which of the necessary properties of a probability sample does this procedure have?

c) For systematic sampling with multiple random starts we might choose 5 random starting points in 1, 2, ..., 500 and then take every 500th entry starting from each of the 5 random points. Which of the necessary properties of a probability sample does this procedure have?

d) If the list were shuffled into random order before a systematic sample was taken, which of the properties would the procedure have?

e) Treating a systematic sample as if it were a simple random sample often gives good results. Why would this be true?

1.6 Why must all the sampling probabilities be non-zero to get a valid population estimate?

1.7 ⋆ Why must all the pairwise probabilities be non-zero to get a valid uncertainty estimate?

1.8   A probability design assumes that people who are sampled will actually be included in the same, rather than refusing. Look up the response rates for the most recent year of BRFSS and NHANES.

1.9   In a telephone study using random-digit dialing, telephone numbers are sampled with equal probability from a list. When a household is recruited, why is it necessary to ask how many telephones are in the household, and what should be done with this information in computing the sampling weights?

1.10   ★ Derive the Horvitz–Thompson variance estimator for the total, as follows
    a) Write $R_i = 1$ if individual i is in the sample, $R_i = 0$ otherwise. Show that $\mathrm{var}[R_i] = \pi_i(1 - \pi_i)$ and that $\mathrm{cov}[R_i, R_j] = \pi_{ij} - \pi_i \pi_j$.
    b) Show that the variance of the Horvitz–Thompson estimator is

$$\mathrm{var}\left[\hat{T}_{HT}\right] = \sum_{i=1}^{N} \sum_{j=1}^{N} \check{x}_i \check{x}_j (\pi_{ij} - \pi_i \pi_j).$$

    c) Show that an unbiased estimator of the variance is

$$\widehat{\mathrm{var}}\left[\hat{T}_{HT}\right] = \sum_{i=1}^{N} \sum_{j=1}^{N} \frac{R_i R_j}{\pi_{ij}} \check{x}_i \check{x}_j (\pi_{ij} - \pi_i \pi_j).$$

    d) Show that the previous expression simplifies to equation 1.2.

1.11   ★ Another popular way to write the Horvitz–Thompson variance estimator is

$$\widehat{\mathrm{var}}\left[\hat{T}_{HT}\right] = \sum_{i=1}^{n} x_i^2 \frac{1 - \pi_i}{\pi_i^2} + \sum_{i \neq j} x_i x_j \frac{\pi_{ij} - \pi_i \pi_j}{\pi_i \pi_j \pi_{ij}}.$$

Show that this is equivalent to equation 1.2.

CHAPTER 2

SIMPLE AND STRATIFIED SAMPLING

In which there are two kinds of people.

2.1 ANALYZING SIMPLE RANDOM SAMPLES

With a simple random sample of size n from a population of size N all the sampling weights are equal to N/n. We saw in the previous chapter that the Horvitz–Thompson estimator of the population total of a variable X is

$$\hat{T}_X = \sum_{i=1}^{n} \check{X}_i = \frac{N}{n} \sum_{i=1}^{n} X_i. \tag{2.1}$$

For useful inference we also need a standard error estimator. The variance of the Horvitz–Thompson estimator can be written as

$$\mathrm{var}\left[\hat{T}_X\right] = \frac{N-n}{N} \times N^2 \times \frac{\mathrm{var}\,[X]}{n} \tag{2.2}$$

Complex Surveys: A Guide to Analysis Using R. By Thomas Lumley
Copyright © 2010 John Wiley & Sons, Inc.

and we can estimate var $[X]$ by the usual formula. In equation 2.2, the last term is the formula for the variance of a mean, and the second term rescales from the mean to the total. The first term will not be familiar from model-based statistics. It is the *finite population correction*, which takes account of the reduction in uncertainty when a large fraction of the population ends up in the sample. The standard error of the total is the square root of var $\left[\hat{T}_X\right]$.

One way to understand the form of the finite population correction is to think about taking repeated independent samples of size n from the same population. The fraction n/N is the expected proportion of overlap between any two samples, and so $(N-n)/N$ is the fraction of the second sample that will be previously unseen observations. The finite population correction thus represents the reduction in variability between samples that happens because different samples include the same people. Another way to view the formula is that there is no uncertainty about the n individuals we have sampled, so that the assumption of simple random sampling is needed only to estimate the uncertainty in the total for the $N-n$ individuals remaining in the population. Exercise 2.10 verifies that 2.2 is in fact the Horvitz–Thompson variance estimator.

It is very common for n to be much smaller than N, so finite population corrections can often be ignored in survey sampling. The finite population correction also disappears from the formula if the sample is taken with replacement, i.e., if the same individual can be sampled more than once. Sampling with replacement is almost never actually used, but designs are often described as "with replacement" when the finite population correction is ignored and the survey is analyzed as if it had sampling with replacement.

The population mean of X can be estimated by dividing the estimated total by the population size, N

$$\hat{\mu}_X = \frac{1}{N}\sum_{i=1}^{n} \check{X}_i = \frac{1}{n}\sum_{i=1}^{n} X_i, \tag{2.3}$$

so the estimate is just the sample average. The variance estimate is obtained by dividing the variance estimate for the total by N^2

$$\widehat{\text{var}}[\hat{\mu}_X] = \frac{N-n}{N} \times \frac{\widehat{\text{var}}[X]}{n}, \tag{2.4}$$

and the standard error of the mean is the square root of var $[\hat{\mu}_X]$. This formula shows that the uncertainty in the mean is not very sensitive to the population size as long as the population is much larger than the sample. A sample of 100 people gives the same uncertainty about the mean of a population of 10,000 or 100,000,000.

Equation 2.3 can be interpreted another way. The population size is the total of a variable that is equal to one for each person in the population, so the Horvitz–Thompson estimator of the population size is

$$\hat{N} = \sum_{i=1}^{n} \frac{1}{\pi_i}.$$

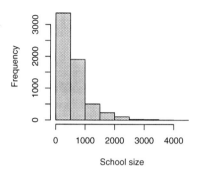 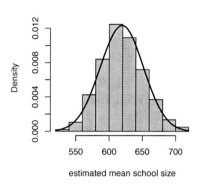

Figure 2.1   Distribution of California school size and of estimated mean school size in samples of 200 schools, with Normal density curve based on the Central Limit Theorem

Since $\pi_i = n/N$ in this sampling design, $\hat{N} = N$, but for other sampling designs this need not be the case. Neither estimator is uniformly more accurate, but the estimator based on $\hat{N}$, the *ratio estimator of the mean* is popular in software, largely because N is not always known. For large surveys where the population size is known, the sampling weights are usually adjusted so that $\hat{N} = N$, as discussed in Chapter 7, so the choice of mean estimator is usually not important.

Equations 2.2 and 2.4 appear to imply that larger sample sizes are always better, a conclusion that would be true if non-response was not a concern or resources were unlimited. In practice, increasing the sample size will reduce the amount of effort that can be expended on follow-up for each individual sampled, increasing the rate of non-response, and introducing biases that can easily outweigh the reduction in variance.

2.1.1   Confidence intervals

Confidence intervals for estimates are computed by using a Normal distribution for the estimate, ie, for a 95% confidence interval adding and subtracting 1.96 standard errors. This is not the same as assuming a Normal distribution for the data. Under simple random sampling and the other sampling designs in this book the distribution of estimates across repeated surveys will be close to a Normal distribution (from the Central Limit Theorem) as long as the sample size is large enough and the estimate is not too strongly influenced by the values of just a few observations.

Figure 2.1 demonstrates this for the mean of school size in California, using the API population data built in to the survey package. The left-hand graph shows the distribution of school size (`apipop$enroll`), the right-hand side shows the distribution of mean school size across 1000 samples of size 200 taken from this population. The estimated means have a distribution that is very close to Normal, centered around the true population mean of 619. The solid curve shows the Normal distribution that would be assumed when estimating a confidence interval, with mean equal to the population mean and variance given by equation 2.2.

2.1.2   Describing the sample to R

Although simple random samples do not require specialized software for analysis, they do provide a simple example for introducting the **survey** package. The first step in analyzing data is to describe the design to R. The svydesign() function takes this description and adds it to the data set to produce a survey design object. The survey design object is then used in all analyses. Wrapping the design information and the data up in a single object ensures that the design information cannot be inadvertently separated or used with the wrong data set.

The **survey** package includes a simple random sample from the API population, in the data set apisrs. The variable fpc in this data set contains the number 6194, the number of schools in California. Figure 2.2 shows the code to describe this survey to R. In the call to svydesign() the data=apisrs argument specifies where the data are stored. The argument id=~1 says that individual schools were sampled, and fpc=~fpc says that the variable called fpc in the data set contains the population size (the ~ notation denotes a variable in the supplied data set). The sampling weights can be worked out from the population size and sample size, so they do not need to be specified. The survey design object has been called srs_design, and when it is printed it will display some basic identifying information. The functions svymean()

```
> library(survey)
> data(api)
> srs_design <- svydesign(id=~1, fpc=~fpc,  data=apisrs)
> srs_design
Independent Sampling design
svydesign(id = ~1, fpc = ~fpc, data = apisrs)
> svytotal(~enroll, srs_design)
         total      SE
enroll 3621074 169520
>  svymean(~enroll, srs_design)
          mean      SE
enroll 584.61 27.368
```

Figure 2.2   Describing a simple random sample to R

and svytotal() estimate the population mean and population total, respectively. As with svydesign(), the ~ notation gives the name of a variable in the survey design object. The estimated population total is 3.6 million students enrolled, with a standard error of 169,000; the estimated mean school size is about 585, with a standard error of 27. The true values are 3.8 million and 619, respectively, so the standard errors do give an accurate representation of the uncertainty in the estimates.

The finite population correction has very little impact on these estimates, and we might safely have ignored it.

If the population size is not specified it is necessary to specify the sampling probabilities or sampling weights. The variable pw in the data set contains the sampling weight, $6194/200 = 30.97$. Figure 2.3 shows the impact of omitting the population size. When the design object is printed, the lack of population size information is indicated by "(with replacement)" in the output. The estimated mean and total are the same, but the standard errors are very slightly larger.

```
> nofpc <- svydesign(id=~1, weights=~pw, data=apisrs)
> nofpc
Independent Sampling design (with replacement)
svydesign(id = ~1, weights = ~pw, data = apisrs)
> svytotal(~enroll, nofpc)
         total      SE
enroll 3621074 172325
> svymean(~enroll, nofpc)
          mean     SE
enroll 584.61 27.821
```

Figure 2.3 Describing a simple random sample without population size information

The svymean() and svytotal() functions can also be applied to a categorical (or "factor") variable. In this case they will create a table of estimated population proportions or estimated population counts for each category of the factor. For example, svytotal(~stype, srs_design) estimates the total number of elementary schools, middle schools, and high schools in California.

Multiple variables can be analyzed in the same call to svytotal() or svymean(), and contrasts can be computed from the results. Figure 2.4 computes the mean of the Academic Perfomance Index for the years 1999 and 2000. The call to svycontrast() computes the difference between the 1999 and 2000 means:

$$(1 \times 2000 \text{ mean}) + (-1 \times 1999 \text{ mean}).$$

An alternative notation is

```
svycontrast(means, quote(api00-api99))
```

where quote() is used to quote an arithmetic expression.

The calls to update() create additional variables in the survey design object. Note that these calls return a new survey design object that must be given a name. In this case it is assigned the same name as the original object.

2.2 STRATIFIED SAMPLING

Simple random samples are not used very often in large surveys, because other designs can give the same precision at lower cost. One way to increase precision is stratified

```
> means <- svymean(~api00+api99, srs_design)
> means
        mean    SE
api00 656.59 9.2497
api99 624.68 9.5003
> svycontrast(means, c(api00=1, api99=-1))
          contrast    SE
contrast     31.9 2.0905
> srs_design <- update(srs_design, apidiff=api00-api99)
> srs_design <- update(srs_design, apipct = apidiff/api99)
> svymean(~apidiff+apipct, srs_design)
            mean     SE
apidiff 31.900000 2.0905
apipct   0.056087 0.0041
```

Figure 2.4   Contrasts and new variables

sampling, which involves dividing the population up into groups called *strata* and drawing a separate probability sample from each one. Stratification ensures that a prespecified number of observations from each stratum end up in the sample, rather than allowing distribution of observations across strata to be random. The sample is less variable, and so gives more precise estimates. The primary limitation to stratified sampling is that, in order to sample within a stratum, the stratum membership must be known for every individual in the population. It is worth noting that the term "stratification" is used in epidemiology for an entirely different process where a sample is analyzed separately in groups based on the values of a variable observed on the sample. Stratified analysis in the epidemiological sense is described in section 2.5.

Since a stratified sample is just a set of simple random samples from each stratum, the Horvitz–Thompson estimator of the total is just the sum of the estimated totals in each stratum and its variance is the sum of the estimated variances in each stratum. The population mean is estimated by dividing the estimated population total by the population size N.

A stratified random sample of 200 schools from the API population is in the apistrat data set. The sample is stratified on school type, with $n_E = 100$ elementary schools, $n_M = 50$ middle schools, and $n_H = 50$ high schools. Since high schools are typically larger than middle schools or elementary schools, a simple random sample that by chance had more high schools would tend to give higher estimated mean and total enrollment, and a sample that by chance had fewer high schools would tend to give lower estimated mean and total enrollment. Fixing the number of schools of each type should decrease the variance.

Figure 2.5 describes this design. In the call to svydesign() that describes this design there is a new option strata=~stype that specifies the stratifying variable. The other change is less visible: in this data set the variable fpc is the population size

```
> strat_design <- svydesign(id=~1, strata=~stype, fpc=~fpc,
    data=apistrat)
> strat_design
Stratified Independent Sampling design
svydesign(id = ~1, strata = ~stype, fpc = ~fpc, data = apistrat)
> svytotal(~enroll, strat_design)
        total      SE
enroll 3687178 114642
> svymean(~enroll, strat_design)
        mean      SE
enroll 595.28 18.509
> svytotal(~stype, strat_design)
       total SE
stypeE  4421  0
stypeH   755  0
stypeM  1018  0
```

Figure 2.5 Describing a stratified design to R

for each stratum, not for the whole population, 4421 for elementary schools, 1018 for middle schools, 755 for high schools.

Stratification has reduced the standard errors by about one third. It is useful to think about where this extra information has come from. Stratifying on school type is possible only if the type of each school in the population is known in advance. This extra information on the population is the source of the extra precision in the estimates, and the precision increase is greatest for the variables that are most accurately predicted by the population information. There is no gain in precision for the Academic Performance Index itself, which has a similar distribution across school types. At the other extreme, the estimated number of schools of each type, computed by svytotal(~stype, strat_design), has no uncertainty, and R reports a zero standard error, since knowing the type of every school in the population means knowing the number of schools of each type.

2.3 REPLICATE WEIGHTS

The standard error of a mean or any other population summary is, by definition, the standard deviation of that estimated summary across many independent samples of data. The replicate-weight approach to estimating standard errors computes the standard deviation of the estimated summary across many partially independent subsets of the one sample, and extrapolates from this to the standard deviation between completely independent samples.

The idea is clearest for split-half replicates. Consider a stratified design that samples two individuals in each stratum. The data can be split into independent half-samples by taking one of the two individuals from each stratum. If the finite-population correction can be ignored, the variance of an estimate based on a half-sample is exactly twice the variance based on the full sample. Since there are two independent half-samples, the variance can be estimated from their difference. Writing $\hat{T}$ for the estimated total and $\hat{T}_A$, $\hat{T}_B$ for the totals estimated from the half-samples

$$E\left[\left(\hat{T}_A - \hat{T}_B\right)^2\right] = 2 \times \text{var}\left[\hat{T}_A\right] = 4 \times \text{var}\left[\hat{T}\right]. \tag{2.5}$$

The estimator based on just one set of splits is unbiased, but very variable; a useful estimator can be obtained by averaging over many sets of splits. With K strata there are 2^K ways to choose one of two individuals from each stratum, which is likely to be an infeasibly large number, so it is necessary to choose a smaller set of splits, preferably in some systematic manner.

It turns out that it is possible to find a set of at most $K + 4$ splits that gives exactly the same variance estimate for a population total as using all 2^K. The necessary properties of the set of splits are that every observation is in $\hat{T}_A$ for half the splits, and that every pair of observations from different strata is in $\hat{T}_A$ for 1/4 of the splits and in $\hat{T}_B$ for 1/4 of the splits. These properties together are called *full orthogonal balance*, and the method is called *Balanced Repeated Replication* or *BRR* [104]. The computation of estimates from half-samples is implemented in practice by assigning a weight of $2/\pi_i$ to the observation that is included and $0/\pi_i$ to the other observation, so a set of replicates is specified by a set of variables each containing these modified weights for one split.

The BRR variance estimator is exactly unbiased and exactly the same as using all 2^K possible replicates for the population total. For other estimated population summaries these properties hold only approximately. Slightly different estimates can be obtained from different sets of replicates in full orthogonal balance, or from different forms of the variance calculation. Appendix C.2 gives some details of how R computes BRR variance estimates.

Fay's method [67] modifies BRR to use all the observations in every replicate, by assigning weights $(2 - \rho)/\pi_i$ and ρ/π_i for some ρ between 0 and 1, with $\rho = 0.3$ being a common choice. This modification reduces the problems that BRR can have with zero cells in 2×2 tables and similar small-sample numerical difficulties.

Splitting a sample into independent subsets is more difficult with other designs, so the replicate weight approaches for other designs use overlapping subsamples. Jackknife estimates [44] set the weight to 0 for a single individual (or cluster: see Chapter 3) and to $n/(n-1)$ for the other observations. This method is often abbreviated "JK1" for an unstratified sample and "JKn" for a stratified sample.

Bootstrap estimates take a sample of observations (or clusters) with replacement from each stratum and the sampling weight is multiplied by the number of times the observation appears in this sample (Rao & Wu [131, 132], Canty & Davison [27]). Details of many variants of these methods are given by Shao and Tu [163] and by Wolter [188].

Replicate weight methods require more computation than using the Horvitz–Thompson estimator but are easier to generalize to statistics other than the mean and total (at least for software designers). For example, correct standard errors for a subpopulation estimate are obtained simply by restricting the estimation to individuals from that subpopulation. The techniques described in Chapter 7 were only available in software using replicate weights until quite recently.

2.3.1   Specifying replicate weights to R

The 2005 California Health Interview Survey [23] provides 80 sets of replicate weights, and instructions for using these weights in Stata or SUDAAN [25]. The data for adults can be downloaded in Stata format, which can be read into R using read.dta() from the foreign package. Column 419 of the data gives the sampling weights, and columns 420–499 give the replicate weights.

The function svrepdesign() creates a survey design object using replicate weights supplied with the data. The sampling weights are supplied as the weights argument and the replicate weights as the repweights argument. These can be model formulas, but it is usually more convenient just to give the appropriate columns of the data set, as in this example:

```
chis_adult <- read.dta("adult.dta")
chis <- svrepdesign(variables=chis_adult[,1:418],
    repweights=chis_adult[,420:499],
    weights=chis_adult[,419], combined.weights=TRUE,
    type="other", scale=1, rscales=1)
```

The option combined.weights specifies that the replicate weights include the sampling weight; the alternative is that they need to be multiplied by the sampling weight.

The scale and rscales options give an overall multiplier and a vector of multipliers for each replicate to rescale the variances. The variances for a statistic $\hat{T}$ based on M replicates $T_1^*, T_2^*, \ldots, T_n^*$ are computed as

$$\widehat{\text{var}}\left[\hat{T}\right] = A \times \sum_{r=1}^{M} a_r \times \left(T_r^* - \bar{T}^*\right)^2$$

where A is scale and a_i are the elements of rscales, and $\bar{T}^*$ is the average of the replicates. Bootstrap and BRR replicates will have $A \approx 1/M$, for jackknife-based replicates $A \approx 1$. For the CHIS A and a_r are all 1; no rescaling is necessary. For some replicate weight methods (JK1, JKn, BRR, Fay's method) R can work out these multipliers from the type argument, but CHIS uses a different method, called JK2.

2.3.2   Creating replicate weights in R

The survey package can also create replicate weights from a design specified by svydesign(), using the function as.svrepdesign(). The default is to create

```
> boot_design <- as.svrepdesign(strat_design, type="bootstrap",
    replicates=100)
> jk_design <- as.svrepdesign(strat_design)
> boot_design
Call: as.svrepdesign(strat_design, type = "bootstrap",
    replicates = 100)
Survey bootstrap with 100 replicates.
> jk_design
Call: as.svrepdesign(strat_design)
Stratified cluster jackknife (JKn) with 200 replicates.
> svymean(~enroll, boot_design)
        mean     SE
enroll 595.28 18.856
> svymean(~enroll, jk_design)
        mean     SE
enroll 595.28 18.509
```

Figure 2.6   Creating replicate weights for a stratified sample

jackknife weights, the `type` option allows for BRR, bootstrap, and Fay's method as well. For example, using the `strat_design` object created in section 2.2 we can create design objects with jackknife (JKn) and bootstrap replicates as shown in Figure 2.6.

The most obvious advantage of designs with replicate weights in the survey package is that it is possible to compute standard errors for differences between subpopulations for arbitrary statistics (see section 2.5). Comparisons of means and proportions, the most common cases, are possible for any design object using regression, as will be discussed in Chapter 5. Comparisons for medians or other summaries are more difficult.

Another reason to create replicate weights from a design is to perform an analysis that has not been implemented in the survey package. Using replicate weights it is only necessary to write code to compute the weighted point estimates; this code is re-run on all the sets of replicate weights to estimate standard errors, which is usually easier than writing code for the linearization standard error estimates. The function `withReplicates()` evaluates a user-supplied function or expression with each set of replicate weights in a design and computes the standard error estimates.

For an example, consider confidence intervals for survival probabilities. Using the NWTS population, we take an artificial sample of all the children who relapsed and 10% of the rest, and estimate the five-year survival probability in the full NWTS cohort. The code is in Figure 2.7. The first six lines of code construct the sample data set. The function `fivesurv()` computes the weighted Kaplan–Meier estimator of survival using the function `survfit()` from the survival package, and then extracts the survival estimate at the first time point after five years.

```
popn <- read.csv("nwts-share.csv")
cases <- subset(popn, relaps==1)
cases$wt <- 1
ctrls <- subset(popn ,relaps==0)
ctrls$wt <- 10
samp <- rbind(cases, ctrls[sample(nrow(ctrls), 325),])

fivesurv<-function(time,status,w){
            scurve <- survfit(Surv(time,status)~1, weights=w)
            scurve$surv[min(which(scurve$time > 5))]
}


des <- svydesign(id=~1, strata=~relaps, weights=~wt, data=samp)
jkdes <- as.svrepdesign(des)
withReplicates(jkdes, quote(fivesurv(trel, relaps, .weights)))
```

Figure 2.7 Replicate weights for five-year survival

The weighted Kaplan–Meier estimator uses the same principle of scaling the data up to the population with sampling weights. The Kaplan–Meier estimator is based on multiplying together the estimated proportions of the sample surviving through short intervals of time, to obtain the probability of surviving for any time interval. The weighted estimator simply replaces the sample proportions for the short time intervals with an estimated population proportion: the total sampling weight for those who survive through the interval, divided by the total sampling weight for those present at the beginning of the interval.

The stratified sampling design is described in the svydesign() call, and then converted to a design with replicate weights. The call to withReplicates has the design object as its first argument, and an expression giving the analysis as the second argument. In this expression the data variables are available by name, and the weights are available as the variable .weights. The resulting estimate is 0.8366 with standard error 0.0016. The very small standard error is because the sample includes all the events from the population. Note that the standard error is for estimating five-year survival in the NWTS cohort, not in the hypothetical superpopulation of all children with Wilms' tumor. For methods for the latter, more interesting estimate, see Chapter 8.

The weights in any survey design object can be extracted with the function weights(), with the option type="sampling" for the sampling weights, and type="analysis" for the replicate weights.

Other notes on creating replicate weights. The bootstrap is straightforward only when all the strata are large. There is a stratum-size correction of $n_k/(n_k - 1)$ used as A in equation 2.5, which corrects the variance when the stratum sample size is small. When the sample sizes are small and not all the same it is not clear what correction to use; as.svrepdesign() bases the correction on the harmonic mean of the stratum sizes. The theory for the bootstrap was developed only for the situation of equal-probability sampling within each stratum, it is not clear how well it would work with arbitrary probabilities.

The jackknife and bootstrap can handle finite-population corrections. These corrections will automatically be included if they were present in the original survey design object.

Although in principle it is possible to construct BRR designs with at most $K + 4$ replicates for K strata, the survey package does not know all of the possible sets of splits and will sometimes use more replicates than necessary. The excess is at most 4 for $K < 180$ and at most 5% for $K > 100$. Appendix C.2 and the help page for hadamard() give references and some details on the constructions that are used.

2.4   OTHER POPULATION SUMMARIES

The population mean and the differences in means estimated in section 2.1 are functions of the population total, so the estimates and their variances can be computed from the estimate and variance of the Horvitz–Thompson estimator. There are many other population statistics that can be estimated in a similar way as explicit functions of population totals or means. Other estimators, such as the median or a regression coefficient, are obtained by solving equations that can be written as population totals or means. Estimates under complex sampling can be obtained by estimating the population equation using the Horvitz–Thompson estimator and then solving it.

Standard errors for these estimates can be obtained with replicate weights simply by repeating the estimation for each set of replicates. Standard errors based on design information use an approach called *linearization* to translate the Horvitz–Thompson standard error estimator for the estimated population equation into a standard error for the estimate. Some technical details are given in Appendix A.2.

2.4.1   Quantiles

The median is an example of a statistic defined implicitly rather than explicitly in terms of population means and totals. The proportion of the population below a threshold is the mean of a variable that is 1 for observations below the threshold and 0 for observations above the threshold. Estimating this proportion for all thresholds gives the estimated population cumulative distribution function. The estimated median is the point where the estimated population distribution function crosses one half: the estimated population size above and below the median are equal. Other quantiles are handled in a similar way, for example, the estimated 90th percentile has 90% of the estimated population size below and 10% above.

Medians and other quantiles present some technical difficulties in estimation, because the median depends less smoothly on the data than most commonly calculated summaries. Even under simple random sampling the median is not unique with a sample of even size: any number between the middle two observations is a valid

```
> svyquantile(~bmi_p, design=chis,quantiles=c(0.25,0.5,0.75))
Statistic:
      bmi_p
q0.25 22.68
q0.5  25.75
q0.75 29.18
SE:
          bmi_p
q0.25 0.03250027
q0.5  0.03749990
q0.75 0.05250025

> svyquantile(~api00, strat_design, c(0.25, 0.5, 0.75),ci=TRUE)
$quantiles
            0.25      0.5      0.75
api00 562.2056 667.2358 755.1226

$CIs
, , api00

            0.25      0.5      0.75
(lower 534.0000 636.4618 725.1052
upper) 594.2887 681.0000 776.9814
```

Figure 2.8   Quartiles with standard errors from replicate weights in CHIS and confidence intervals from design information in the API population

choice. svyquantile() interpolates linearly between the two adjacent observations when the quantile is not uniquely defined. Another issue arises with tied or discrete data, in deciding whether two observations with the same value are handled the same as one observation with a larger weight. The details of handling discrete data will only matter when the standard error of the estimated median is comparable to the accuracy with which the data are recorded. Appendix C.4 gives more details.

There are also difficulties in computing confidence intervals for the median. The approach used in survey statistics, and also in survival analysis, is to create a confidence interval for the cumulative distribution function, and use the values where this interval crosses 50% to form a confidence interval for the median. The same approach works with other quantiles, although reliable estimation becomes increasingly difficult for extreme quantiles. This approach originated with Woodruff [189]. Dorfman and Valliant [42] compare some variations on the method.

For objects with design information, created by svydesign(), the uncertainty is reported as the confidence interval. For objects with replicate weights a standard error is reported. Figure 2.8 shows the use of svyquantile() for the strat_design

```
> tab<-svymean(~interaction(ins,smoking,drop=TRUE),chis)
> tab
[...output edited to fit...]
                                     mean       SE
YES.CURRENTLY SMOKES              0.112159 0.0021
NO.CURRENTLY SMOKES               0.039402 0.0015
YES.QUIT SMOKING                  0.218909 0.0026
NO.QUIT SMOKING                   0.026470 0.0012
YES.NEVER SMOKED REGULARLY 0.507728 0.0036
NO.NEVER SMOKED REGULARLY   0.095332 0.0022
> ftab<-ftable(tab,rownames=list(ins=c("Insured","Uninsured"),
    smoking=c("Current","Former","Never")))
> ftab
                ins      Insured    Uninsured
smoking
Current mean         0.112158606 0.039402123
        SE           0.002125764 0.001487859
Former  mean         0.218908955 0.026470374
        SE           0.002600444 0.001206811
Never   mean         0.507728323 0.095331620
        SE           0.003614234 0.002210343
> round(ftab*100,1)
                ins Insured Uninsured
smoking
Current mean          11.2       3.9
        SE             0.2       0.1
Former  mean          21.9       2.6
        SE             0.3       0.1
Never   mean          50.8       9.5
        SE             0.4       0.2
```

Figure 2.9   Smoking by insurance status in the California Health Interview Survey

sample of California schools and for the replicate-weight analysis of the California Health Interview Survey.

2.4.2 Contingency tables

Two-way or multi-way tables with standard errors for the counts in each cell can be produced with svytotal() using the function interaction() to make a factor variable with a level for each cell of the table. For example, Figure 2.9 shows a table of smoking by insurance status with data from the California Health Interview Survey described in section 2.3.1. The initial output has one row for each cell of the table. Reformatting the output to look like a two-way table is possible with the ftable()

function. The `rownames` argument to `ftable()` specify the names to be used for each level of the two variables. The resulting table is a matrix of numbers that can be multiplied by 100 to give percentages, and then rounded. We see that current smokers are the most likely to be uninsured. Former smokers are the least likely to be uninsured. This might be an effect of age, since older people are more likely to be insured and more likely to be former smokers, but the pattern is still visible in the subpopulation younger than 55

Testing for association in a two-way table presents some complications for design-based influence. A p-value in a hypothesis test is computed using the distribution of the test statistic when the null hypothesis is true. That is, the null-hypothesis distribution is based on sampling from a population that may be quite dissimilar from the actual population. It is not obvious how this distribution should be generated without making modelling assumptions.

This statistical complication is related to the substantive question of what the null hypothesis should be. In a fixed finite population the population-level association between two variables is either exactly zero, or, much more likely, not exactly zero. It is not clear that anyone should be interested in knowing which of these is true. For example, if there are 56,181,887 women and 62,710,561 men in a population it is not possible for the proportions of men and women who are unemployed to be the same, since these population sizes have no common factors. We would know without collecting any employment data that the finite-population null hypothesis was false.

A more interesting question is whether the finite population could have arisen from a process that had no association between the variables: is the difference at the population level small enough to have arisen by chance. Technically this is a two-phase sampling question, considered in Chapter 8. A much simpler approach to the question is to treat the sample as if it came from an infinite *superpopulation*, and simply ignore the finite-population corrections in inference.

Under this superpopulation sampling the large-sample distribution of the Pearson χ^2 statistic was computed by Rao and Scott [130]. The distribution is not a standard one, so Rao and Scott also gave number of approximations. The first approximation corrects the mean of the statistic, the second also corrects the variance. These are available from the function `svychisq()` using `statistic="Chisq"` and `statistic="F"`. Another approach to testing for association is to estimate a set of association parameters and test whether they are zero. These tests were proposed by Koch et al. [77] and are available using `statistic="Wald"` and `statistic="adjWald"` in `svychisq()`.

Based on simulations by Thomas and Rao[169], the second-order Rao–Scott approximation is the most accurate of these commonly-used tests and `svychisq()` uses that as the default. The exact large-sample distribution was not examined by Thomas and Rao. Two approaches to computing it are implemented in `svychisq()` and discussed in Appendix A.3. This should be more accurate but is not the default because it is not yet widely used and its performance in small samples has not been studied, and because it takes too much computing time when the p-value is very small.

In the CHIS example in Figure 2.9 the test for independence of smoking and insurance is requested by svychisq(~smoking+ins, chis). The p-value is extremely small. The exact large-sample p-value cannot be computed and the other four approximations all agree that $p < 2.2 \times 10^{-16}$.

Log-linear models for more detailed analysis of contingency tables are discussed in section 6.3.

2.5 ESTIMATES IN SUBPOPULATIONS

Since each stratum of a stratified design is a separate simple random sample it is easy to provide estimates of means, totals, and other statistics for each stratum. For example, the BRFSS is stratified by state, and state-specific estimates of behavioral risk factors will often be of interest. Estimates for a single state can be obtained by using only the data from that state, and specifying the design for that stratum.

For subpopulations that are not strata, the situation is more complicated. For example, one might want to have race-specific or age-specific estimates of behavioral risk factors from BRFSS, but these cannot be obtained just by ignoring the data from other race or age groups. For example, the 18–25 year age group in BRFSS is not a stratum, and the pairwise sampling probabilities π_{ij} are not the same as they would be if this group were a stratum, because the number of participants from this age group is random rather than fixed in advance. Since the sampling weights would be correct but the pairwise sampling probability would be incorrect, the resulting point estimates would be right, but the standard errors would be wrong. The problem of estimation in subpopulations is also called *domain estimation* in the survey literature.

Fortunately, the survey package handles these details without any special effort from the user. As long as the full sample is used to define the survey design object, estimates can be computed for any subpopulation of interest (see Appendix A.2.1 for details of how this is done). There are two ways to specify estimates on a subpopulation. The first is to create a survey design object that represents the subpopulation, using the subset() function.

The emer variable in the API population is the percentage of teachers with only emergency teaching certification, an indicator of difficulty in recruiting staff to the school. About 20% of schools have no teachers with emergency certification and about the same number have more than 20% of teachers with emergency certification. We can estimate the mean Academic Performance Index and the total number of students in both subsets, as shown in Figure 2.10

The second approach is to use svyby() to compute estimates for a set of subpopulations. With the CHIS survey design object defined in section 2.3.1, the following code computes mean BMI for sex and race/ethnicity subpopulations, giving the output in Figure 2.11. The first argument to svyby() specifies the variables to be analyzed, the second specifies the grouping variables that define the subpopulations, and the third specifies the analysis to be performed on each subpopulation. In this example the variable to be analyzed is BMI (bmi_p), the grouping variables are sex and race (srsex and racehpr), and the requested analysis is a mean (svymean).

```
> emerg_high <- subset(strat_design, emer>20)
> emerg_low <- subset(strat_design, emer==0)
> svymean(~api00+api99, emerg_high)
        mean    SE
api00 558.52 21.708
api99 523.99 21.584
> svymean(~api00+api99, emerg_low)
        mean    SE
api00 749.09 17.516
api99 720.07 19.061
> svytotal(~enroll, emerg_high)
        total    SE
enroll 762132 128674
> svytotal(~enroll, emerg_low)
        total    SE
enroll 461690 75813
```

Figure 2.10 Estimating in subpopulations

	srsex	racehpr	statistics.bmi_p	se
1	MALE	LATINO	28.2	0.1447
2	FEMALE	LATINO	27.5	0.1443
3	MALE	PACIFIC ISLANDER	29.7	0.7055
4	FEMALE	PACIFIC ISLANDER	27.8	0.9746
5	MALE	AMERICAN INDIAN/ALASKAN NATIVE	28.8	0.5461
6	FEMALE	AMERICAN INDIAN/ALASKAN NATIVE	27.0	0.4212
7	MALE	ASIAN	24.9	0.1406
8	FEMALE	ASIAN	23.0	0.1112
9	MALE	AFRICAN AMERICAN	28.0	0.2663
10	FEMALE	AFRICAN AMERICAN	28.4	0.2417
11	MALE	WHITE	27.0	0.0598
12	FEMALE	WHITE	25.6	0.0680
13	MALE	OTHER SINGLE/MULTIPLE RACE	26.9	0.3742
14	FEMALE	OTHER SINGLE/MULTIPLE RACE	26.7	0.3158

Figure 2.11 Body mass index by race and sex, CHIS 2005

```
bys <- svyby(~bmi_p, ~srsex+racehpr,svymean, design=chis,
    keep.names=FALSE)
print(bys, digits=3)
```

In a design with replicate weights, comparisons between subpopulations can be done for arbitrary statistics. Figure 2.12 shows the estimation of ratios of median BMI for sex and race/ethnicity groups in the California Health Interview Survey. The first line estimates the median for each subpopulation, with the option covmat=TRUE asking for the covariance matrix to be computed. The two calls to svycontrast() compute the ratio of these estimated medians for male vs female Latinos and male Latinos vs male whites. Both ratios are above 1 by a statistically significant amount, but not necessarily a meaningful amount for public health. Ratios are discussed further in section 5.1.

```
> medians <- svyby(~bmi_p, ~srsex+racehpr,svyquantile,
    design=chis, covmat=TRUE,quantiles=0.5)
> svycontrast(medians, quote(MALE.LATINO/FEMALE.LATINO))
         nlcon     SE
contrast 1.0258 0.0076
> svycontrast(medians, quote(MALE.LATINO/MALE.WHITE))
         nlcon     SE
contrast 1.0441 0.0057
```

Figure 2.12   Contrasts between subpopulations using replicate weights

Missing data.   As is usual in R, most estimates based on missing (NA) data give NA results by default. Functions that default to NA in the presence of missing data can be given an option na.rm=TRUE to omit the missing data. The effect of na.rm=TRUE is to treat the non-missing data as a subpopulation.

2.6   DESIGN OF STRATIFIED SAMPLES

As stratification decreases standard errors it would theoretically be ideal to have as many strata as possible, if enough informative population-level data were available to create the strata. In order for all pairwise sampling probabilities to be non-zero, which is required for estimating standard errors correctly, the sample size in a stratum has to be at least two, unless there is only one individual in the population stratum, in which case a sample size of one is permissible. It is important to note that there is nothing unreasonable about having a stratum in which the sampling probability is 100%, for example, some information may be available on all large business but requiring sampling to obtain for small businesses. In a stratum with 100% sampling there is no uncertainty about the population and so no contribution to the variance.

If instead there is a relatively small number of strata the question of the best allocation of sample size to strata arises. The best allocation depends on why the data are being collected. If the goal is to estimate a total across the whole population the formula for the variance of a total (equation 2.2) can be used to gain insights about optimal allocation. The variance of the estimated total is the sum of the variances of estimated totals for the separate simple random sample in each stratum. The variance will be higher for a large stratum (large N^2) or for a heterogenous stratum (large var $[X]$), so it seems reasonable that a larger sample size should be allocated to these strata. It also seems plausible that if all the stratum population sizes were doubled or halved, keeping the same proportions, that the optimal proportions for the sample would not change. This would suggest an allocation proportional to the square root of N^2 var $[X]$ for the stratum (rather than directly proportional to N^2 var $[X]$). Writing n_k for the sample size in stratum k, N_k for the population size in stratum k, and σ_k^2 for the variance of X in stratum k

$$ n_k \propto \sqrt{N_k^2 \sigma_k^2} = N_k \sigma_k. \tag{2.6} $$

Equation 2.6 does turn out to be the theoretically optimal allocation, called Neyman allocation after its inventor [120]. Exercises 2.8 and 2.9 verify the optimality.

In practice it may be difficult to make use of this optimal allocation, especially in a survey carried out for multiple purposes. The stratum variances will have different patterns for different variables, and will often not be known accurately in advance. Even more importantly, estimates may be needed for subsets of the population as well as for the whole population. In order to obtain sufficiently accurate estimates in small strata it is necessary to increase the sample size in those strata beyond what would be optimal for estimating the population total. A similar argument applies for subsets that are not strata, for example, to get good estimates for Hispanics and African–Americans, the NHANES and NHIS surveys sample more heavily from geographic strata that are known (from census data) to have larger minority populations.

Sometimes there are important differences in cost for sampling in different strata, and equation 2.6 can be modified to

$$ n_k \propto \sqrt{\frac{N_k^2 \sigma_k^2}{\text{cost}_k}} = \frac{N_k \sigma_k}{\sqrt{\text{cost}_k}}. \tag{2.7} $$

An example is given by the Behavioral Risk Factor Surveillance System, which carries out telephone surveys using random-digit dialing. The valid telephone numbers in each state are stratified according to whether they appear in the telephone directory. All numbers in the directory correspond to actual telephones and about 40% of these are households. The majority of numbers not in the directory are not in use and so the proportion of households is lower. Since more calls need to be made to reach a household in the unlisted stratum the cost per call will be lower but the cost per household will be higher. The BRFSS Operational Guide [31] suggests 50% higher sampling for listed numbers but says

> States are encouraged to develop a method of determining the actual cost of completing interviews in both high and medium density strata.

EXERCISES

2.1   You are conducting a survey of emergency preparedness at a large HMO, where you want to estimate what proportion of medical staff would be able to get to work after an earthquake.

    a) You can either send out a single questionnaire to all staff, or send out a questionnaire to about 10% of the staff and make follow-up phone calls for questionnaires that are not returned. What are disadvantages of each approach?

    b) You choose to survey just a sample. What would be useful variables to stratify the sampling, and why?

    c) The survey was conducted with just two strata: physicians and other staff. The HMO has 900 physicians and 9000 other staff. You sample 450 physicians and 450 other staff. What are the sampling probabilities in each stratum?

    d) 300 physicians and 150 other staff say they would be able to get to work after an earthquake. Give unbiased estimates of the proportion in each stratum and the total proportion.

    e) How would you explain to the managers that commissioned the study how the estimate was computed and why it wasn't just the number who said "yes" divided by the total number surveyed?

2.2   You are conducting a survey of commuting time and means of transport for a large university. What variables are likely to be available and useful for stratifying the sampling?

2.3   Using the CHIS data, estimate the number of Californians with diagnosed hypertension (variable AB29) and treated hypertension (variable AB30). Do the proportions with these two conditions differ by insurance status (INS), race (RACECEN), or marital status (MARIT).

2.4   Using the CHIS data, estimate the proportion of diagnosed hypertension that is treated (AB29, AB30), the proportion of diabetics (AB22), with diagnosed hypertension, and the proportion of hypertensive diabetics whose diabetes is being treated.

2.5   In the Academic Performance Index data we saw large gains in precision from stratification on school type when estimating mean or total school size, and no gain when estimating mean Academic Performance Index. Would you expect a large or small gain from the following variables: mobility, emer, meals, pcttest? Compare your expectations with the actual results.

2.6   For estimating total school enrollment in the Academic Performance Index population, what is the optimal allocation of a total sample size of 200 stratified by school size? Draw a sample with this optimal allocation and compare the standard errors to the stratified sample in Figure 2.5 for: total enrollment, mean 2000 API, mean meals, mean ell.

2.7 ⋆ Figure 2.1 shows that the mean school size (`enroll`) in simple random samples of size 200 from the Academic Performance Index population has close to a Normal distribution.

    a) Construct similar graphs for simple random samples of size 100, 50, 25, 10.

    b) Repeat for median school size instead of mean school size.

    c) Repeat for mean school size in stratified samples of size 100, 52, 24, 12 using the same stratification proportions (50% elementary schools, 25% middle schools, 25% high schools) as in the built-in stratified sample. [An R function for stratified sampling is on the web page.]

2.8 ⋆ In a design with just two strata write the sample sizes as n_1 and $n - n_1$ so that there is only one quantity that can be varied. Differentiate the variance of the total with respect to n_1 to find the optimal allocation for two strata. Extend this to any number of strata by using the fact that an optimal allocation cannot be improved by moving samples from one stratum to another stratum

2.9 ⋆ Write an R function that takes inputs n_1, n_2, N_1, N_2, σ_1^2, σ_2^2 and computes the variance of the population total in a stratified sample. Choose some reasonable values of the population sizes and variances, and graph this function as n_1 and n_2 change, to find the optimum and to examine how sensitive the variance is the precise values of n_1 and n_2.

2.10 ⋆ Verify that equation 2.2 gives the Horvitz–Thompson estimator of variance for a simple random sample

    a) Show that when $i \neq j$

$$\pi_{ij} = \frac{n}{N}n - 1N - 1.$$

    b) Compute $\pi_{ij} - \pi_i\pi_j$.

    c) Show that the equation in exercise 1.10(c) reduces to equation 2.2.

d) Suppose instead each individual in the population is independently sampled with probability n/N, so that the sample size n is not fixed. Show that the finite population correction disappears from equation 2.2 for this *Bernoulli sampling* design.

CHAPTER 3

CLUSTER SAMPLING

In which birds of a feather flock together.

3.1 INTRODUCTION

3.1.1 Why clusters: the NHANES II design

Stratified random sampling of individuals is feasible for telephone interviews or mail questionnaires, but for in-person interviews it has the problem that the sample is spread out over the entire population. Consider the NHANES examination process. This involves a detailed clinical examination, and blood samples that must be stored and processed for analysis. The National Center for Health Statistics sets up mobile examination centers in large trailers that can be moved around the country, from one site to another.

NHANES II sampled 27,000 people over four years. If the 27,000 people had been a stratified sample of individuals, the mobile examination center would have had to be moved to thousands of locations, with transportation taking up most of the

Complex Surveys: A Guide to Analysis Using R. By Thomas Lumley
39

Figure 3.1   A simple random sample of 10,000 voter locations from the USA. Circles are at the center of each county, with area proportional to the number sampled. The largest sample is 257, in Los Angeles County. County-level data for some areas of New England were not available.

available time and money. Figure 3.1 shows what a sample of 10,000 voters from the US would look like. The sampled individuals live in 1184 different counties, with the largest number from a single country being 257 in Los Angeles County.

In the design of NHANES II, it was felt that logistical considerations precluded having more than 80 locations for the examinations [116]. A sample of size 80 is clearly inadequate, so each sampled location needed to recruit many participants. The final design involved sampling 64 locations and planning to recruit 440 participants in each location. The same principle was then applied within each location: each city or county that was sampled was stratified into census enumeration districts, and groups of households called "segments" were sampled from each district, oversampling districts with a higher Hispanic population.

Cluster sampling, i.e., sampling a relatively small number of groups of people, is almost universal in large surveys that involve in-person interviews. In contrast to stratified sampling, which gives increased precision for the same sample size, cluster sampling decreases precision for a specified sample size, but can increase sample size and precision for a specified cost. The reason that precision is decreased for the same sample size is that people in a cluster tend to be similar: interviewing yet another person in the same town is less informative than interviewing someone from a different town.

3.1.2   Single-stage and multistage designs

Cluster sampling can be used in a single-stage design where the whole population of a cluster is recruited. For clusters such as classrooms, medical practices, or workplaces, it may be more convenient to sample the entire cluster than to attempt to subsample. More commonly, though, sampling of clusters is followed by subsampling within each cluster. Clusters at the first stage are called *Primary Sampling Units* or PSUs, the terms "secondary sampling unit" (SSU), "tertiary sampling unit" for subsequent stages follow logically but are not as widely used.

The overall sampling probability of an element is the product of the sampling probabilities at each stage: if a cluster is sampled with probability 1/100 and a household is sampled from that cluster with probability 1/1000, the sampling probability for the household is $\pi_i = 1/100 \times 1/1000$, and the sampling weight is $100 \times 1000 = 100000$. Some technical conditions on the sampling design are needed for probabilities to work out in this simple way. The two most important are that every individual must be in one and only one cluster, and that the subsampling probabilities for any given cluster do not depend on which other clusters were sampled.

It is also possible to mix cluster sampling and sampling of individuals. Since each stratum of a survey can be thought of as a separate and independent sample, there is no difficulty in having single-stage sampling in one stratum and multistage sampling in another. For example, the Scottish Household Survey stratifies Scotland into high population density and low population density regions. A stratified random sample of individual households is taken in the high-density regions, and a cluster sample is taken in low-density regions where random sampling of individual households would require too much interviewer travel. The sampling probability for a household sampled in a single stage in Glasgow is just the number of households sampled in the stratum, divided by the number in the population stratum, n_h/N_h. The sampling probability for a household in Aberdeenshire, sampled in two stages, is the product of the sampling probability for the cluster and the probability for the household within that cluster.

From a mathematical and computational viewpoint this *multistage sampling* can be most easily analyzed by considering each sampled cluster as a population for further sampling: the variance of an estimate is the sum of variances from each stage of sampling. For example, consider a simplified version of the NHANES II design we first choose 64 regions in 32 strata and then take a simple random sample of 440 individuals within each region. The variance of an estimated total from this design can be partitioned across two sources: the variance of each estimated regional total around the true total for the region, and the variance that would result if the true total for each of the 64 sampled regions were known exactly. The first of these variance components can be estimated directly from equation 2.2, the second can be estimated from a slightly modified version of the same equation. The sequential view of multistage sampling as subsampling and adding extra variance at each stage is computationally simpler than directly using the Horvitz–Thompson formula, and allows any number of stages of sampling. It has been used in software at least since David Bellhouse's program TREES [7] in 1983.

If clusters were sampled with replacement at the first stage, the variance computations would depend on the weights from all stages of sampling but on cluster membership only at the first stage. A sampling design could be specified by giving three variable: the sampling weights, the strata identifiers at stage one, and the PSU identifiers. Since most large surveys sample only a small proportion of the population we would expect there to be little change in standard errors if we pretended that the first stage of sampling was with replacement, and this turns out to be correct. In addition, except in a few artificial examples, pretending that the first stage of sampling was with replacement gives a conservative standard error estimate. When some large clusters are sampled with probability 1 the first stage of sampling contributes no uncertainty for these clusters. In the single-stage approximation the PSUs are treated as strata and the second-stage sampling units are treated as PSUs.

It is very common for large surveys to report only the cluster and strata information for the first stage of sampling, or even to report "pseudo" PSU and strata information for the first stage that gives good approximations to the correct standard errors but does not correspond exactly to the sampling design. In addition to computational simplicity, this practice makes it harder to identify individuals from the survey data, which might well be possible if the full multistage strata and cluster information was published. The only large surveys I have found with publicly available design information beyond stage one are a few years of the National Health Interview Survey. The full multistage design information is still used in internal analyses by the agencies that perform the surveys, but it is not available for secondary analysis.

3.2   DESCRIBING MULTISTAGE DESIGNS TO R

For a single-stage cluster sample, or a multistage sample that is being treated as a single-stage sample with replacement, the only difference in the svydesign() call is that id=~1 is replaced by a formulate that specifies the PSU identifier. For the NHANES III data this is

```
svydesign(id=~SDPPSU6, strat=~SDPSTRA6,  weight=~WTPFQX6,
    nest=TRUE, data=nhanes3)
```

where SDPPSU6 and SDPSTRA6 are the "pseudo"-PSU and stratum identifiers defined for the single-stage analysis. The argument nest=TRUE is used to indicate that the PSU identifier needs to be interpreted as nested within stratum: the same PSU id is recycled in different strata. Without nest=TRUE, svydesign() will check that PSUs do not overlap more than one stratum and report an error if they do. If the design includes finite population correction information, the population size is the number of sampling units in the stratum. For example, in a design that takes a simple random sample of 15 school districts from the API population and observes all the schools in those districts, the population size is 757, the number of school districts in California.

In a multistage sample that uses information about finite population sizes, the id, strata, and fpc arguments must specify a variable for each stage of sampling. If some stages are unstratified and others are stratified, define a variable with a single value as the stratum identifier for the unstratified stage(s). The population size for the fpc argument is the size of the population being sampled: the number of next-stage sampling units in the current cluster or stratum within cluster. For example, consider a two-stage design for the API population that samples 40 school districts, then five schools within each district (or all the schools, if there are fewer than five). The design has population size 757 at the first stage, for the number of school districts in California. At the second stage the sampling is of schools within a district, so the population size is the number of schools in the district. Each argument takes a model formula with two terms, one for each stage of sampling. As with sampling of individuals, the weights need not be specified if they can be worked out from the population sizes.

```
clus2_design <- svydesign(id=~dnum+snum, fpc=~fpc1+fpc2,
    data=apiclus2)
```

3.2.1   Strata with only one PSU

When a population stratum has only one potential PSU the sampling fraction for this stratum must be 100%, since otherwise it would be 0%. The stratum then does not contribute to the first stage variance, but may contribute to variances at later stages of sampling. It is helpful in this situation to ignore the first stage of sampling in this stratum and treat the second stage, sampling within the single PSU, as the first stage for the purposes of variance calculation. Ignoring the first stage of sampling is particularly useful when a single-stage approximation to the variance is being used.

If the population stratum has more than one potential PSU but only one is sampled, the design violates the conditions given in section 1.1.2. If two individuals in the stratum are in different PSUs they cannot both be sampled, so the pairwise sampling probabilities π_{ij} are zero for some pairs. Strata with a single sampled PSU do occur in surveys, for two reasons. The first reason is nonresponse or other problems in carrying out the intended design, the second is a deliberate attempt to reduce variance by fine stratification.

The best way to handle a stratum with only one PSU is to combine it with another stratum, one that is chosen to be similar based on population data available before the study was done. The variance estimates will then be conservative: population estimates from the survey will be more accurate than they would have been if the strata had been combined at the design stage. Some NHANES studies use this approach, taking one PSU per stratum and then creating pseudo-strata with two PSUs for analysis. It is vital that the choice of strata to combine depends on population data rather than sample data: combining strata based on the similarity of sample data will lead to anti-conservative variance estimates.

When strata with a single PSU cannot be (or at least have not been) combined to fix the problem, the default behavior of the **survey** package is to report an error, but it also provides two approximations to fix the problem. These are controlled by options(survey.lonely.psu). Setting options(survey.lonely.psu =

"adjust") gives a conservative variance estimator that uses residuals from the population mean rather than from the stratum mean, and options(survey.lonely.psu = "average") sets the variance contribution to the average for all strata with more than one PSU. The adjust option is conservative. The average option is intended for designs where the lonely PSUs result from nonresponse and where strata are roughly comparable, e.g., a design with two individuals sampled per stratum. Strata with only a single PSU in a subpopulation are likely to lead to poor variance estimation, and a warning is given. The same adjustments are applied at each level of sampling, e.g., to second-stage strata with only a single SSU.

When there is only a single population PSU in a stratum it should be clear from the finite population correction information that the sampling fraction is 100%. If, for some reason, population size information is not supplied, single-PSU strata can be dropped from the variance calculation with options(survey.lonely.psu = "remove").

3.2.2   How good is the single-stage approximation?

One of the few publicly-available data sets that includes multistage sampling information is the 1992 National Health Interview Survey. The finite-population information is still fairly incomplete, specifying just that some *self-representing* population strata had only a single PSU, which was then sampled with probability 1. The NCHS guidelines for analysis [117] recommend that the population size be represented as infinite for all stage-two sampling and as either 1 or infinite in stage-one sampling. The alternative analysis, for software that cannot handle two-stage designs, is a single-stage analysis without finite population correction, using "pseudo" PSU and stratum information.

Figure 3.2 shows the code for declaring the designs and performing one analysis. The finite population corrections for the two-stage design are specified as sampling fractions n/N, not as population sizes N, because the population size is infinite. For finite population sizes either form of the correction is allowed, but they cannot be mixed in the same call to svydesign(). In neither design can the sampling weights be computed from the population sizes, so they are specified in the weights argument.

A warning is given when the two-stage design is defined, indicating that an implausibly large population size has been given. In this case the infinite population size is intended, and the warning can be ignored. The final two lines request estimated means for height (in inches), weight (in pounds), number of acute medical conditions, and total number of medical conditions.

The two-stage analysis gives the same estimated means as the one-stage analysis, because the sampling weights are identical, but gives slightly smaller (and more accurate) standard errors. The difference in the estimated standard errors is about 10%, and in large public-use data sets seems reasonable to accept an inflation in standard errors by 10% in return for the added simplicity and improved confidentiality of the single-stage approximation.

```
> nhis$selfrep<-ifelse(nhis$psutype=="MSA- self-representing",
          1,0)
> nhis$zero<-rep(0,nrow(nhis))
> twostage<-svydesign(id = ~psupseud + segnum,
    strata = ~stratum + subtype,
    weight = ~wtfa, data = nhis,
    fpc = ~selfrep+zero)
Warning message:
FPC implies population larger than ten billion.
    in: as.fpc(fpc, strata, ids)
>
> nhis$cstratum<-nhis$psupseud %/% 10
> nhis$cpsu<-nhis$psupseud %% 10
> onestage<-svydesign(id = ~cpsu, strata = ~cstratum,
    weight = ~wtfa, data = nhis, nest=TRUE)

> svymean(~height+weight+nacute+ncond, onestage, na.rm=TRUE)
          mean       SE
height   67.306288 0.0253
weight  161.112444 0.2017
nacute    0.063934 0.0015
ncond     0.604453 0.0083
>
> svymean(~height+weight+nacute+ncond, twostage, na.rm=TRUE)
          mean       SE
height   67.306288 0.0235
weight  161.112444 0.1879
nacute    0.063934 0.0014
ncond     0.604453 0.0069
```

Figure 3.2 Single-stage and two-stage approximations to the NHIS 1992 design

3.2.3 Replicate weights for multistage samples

Replicate weights for single-stage cluster sampling designs are produced by treating the clusters as units and applying the same methods as for sampling of individuals (section 2.3). The BRR method now applies to designs with two PSUs in each stratum, and creates half-samples with one PSU from each stratum. The jackknife methods set the weight to zero for each PSU in turn. The bootstrap resamples PSUs rather than individuals.

For multistage designs with small sampling fractions the same single-stage without-replacement approximation can be used as for linearization methods. For multistage design with large sampling fractions there are fewer options, one is presented by Funaoka et al. [51] but is not currently implemented in the survey package.

Using replicate weights provided by the study designers works the same way for multistage as single-stage samples: the replicate weights and the scaling values are simply provided in the call to svrepdesign(). When creating replicate weights from a survey design object using as.svrepdesign() the design is treated as single-stage and the jackknife, BRR, or bootstrap procedures are applied to the PSUs to create replicate weights.

3.3 SAMPLING BY SIZE

An old primary-school joke asks "Why do white sheep eat more than black sheep?"; the answer being "There are more white sheep than black sheep."

The fact that totals for any variable tend to increase with population size means that the best cluster-sampled design for estimating one variable is likely to be good for other variables. The strength of this effect is shown in Figure 3.3, which is based on county-level voting data for George W. Bush, John Kerry, and Ralph Nader in the US presidential elections of 2004. The proportions voting for Bush and Kerry are almost perfectly negatively correlated, but the totals voting for the two candidates have a positive correlation of 0.88.

In unstratified sampling the variance of the estimated total of X depends on the sample variance of X. If π_i could be chosen proportional to X_i, this variance would be zero. This would require X_i to be known (up to a scaling factor) for the whole population, so it is not surprising that accurate estimation is possible. More realistically, if π_i is approximately proportional to X_i the variance of the estimated total will be small.

As Figure 3.3 indicates, this is possible for cluster sampling, since choosing π_i proportional to the population size in the cluster will make it approximately proportional to the cluster total for a wide range of variables. *Probability-proportional-to-size* (PPS) cluster sampling is sufficiently important in survey design that PPS is often used as a general term for any design that is not multistage stratified random sampling. Tillé [172] gives descriptions of the creation and use of PPS designs; an influential earlier reference is Brewer and Hanif [22].

Unfortunately, analysis of PPS sampling without replacement requires knowing the pairwise sampling probabilities π_{ij}. These are usually not supplied in a data set

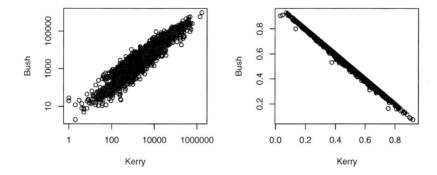

Figure 3.3   Totals (log scale) and proportions by county of the vote for George W. Bush and John Kerry in the 2004 US presidential elections

and computing them requires knowing the size for all clusters in the population, not just the sampled ones, and knowing what algorithm was used to draw the samples. At the time of writing, the survey package can only use the single-stage with-replacement approximation to analyze PPS designs.

Using the sampling package [173] and Tillé's splitting method for PPS sampling without replacement we can draw a sample of 40 counties from the population of 4600 with probability proportional to the total number of votes cast. Figure 3.4 shows the code and results.

```
> election$votes<-with(election, Bush+Kerry+Nader)
> election$p<-40*election$votes/sum(election$votes)
> library(sampling)
> insample<-UPtille(election$p)
> ppsample <- election[insample==1,]
> ppsample$wt<-1/ppsample$p
> pps_design<-svydesign(id=~1, weight=~wt, data=ppsample)
> svytotal(~Bush+Kerry+Nader, pps_design,deff=TRUE)
          total      SE    DEff
Bush   64518472  2671455 0.0028
Kerry  51202102  2679433 0.0016
Nader    478530   105303 0.0756
> colSums(election[,4:6])
    Bush     Kerry    Nader
59645156  56149771   404178
```

Figure 3.4   Estimating the total presidential votes from a PPS sample

In this example we can compute an approximation (Hartley and Rao [59]) to the joint sampling probabilities π_{ij} and Horvitz–Thompson standard error estimator using the known sizes of all the counties (the code is not shown, but is on the website). This gives standard errors of 2,620,000 for Bush, 2,530,000 for Kerry and 103,000 for Nader. We can also compute the true joint sampling probabilities, using the function UPtillepi2() in the **sampling** package. The Horvitz–Thompson estimator based on the true joint probabilities gives standard errors of 2,600,000 for Bush, 2,520,000 for Kerry, and 102,000 for Nader. The Hartley–Rao approximation is not meaningfully different from the result with the true joint probabilities, and takes seconds rather than hours to compute. The single-stage with-replacement approximation appears to be over-estimating the standard errors by about 6% for the major candidates and about 2.5% for Nader. The effect of the finite population size is larger than it would be for a simple random sample of the same size, because some of the sampling probabilities are quite large: Los Angeles County has $\pi_i = 0.9$.

It is important to remember that these are only estimates from a single sample. A real evaluation of the impact of the single-stage with-replacement approximation would require a simulation study, such as Exercise 3.2. Simulation does lead to the same conclusion: that the single-stage approximation is acceptably accurate, although less accurate than it is in a stratified multistage design.

The Family Resources Survey () uses a PPS sample of postcode sectors at the first of its three stages of sampling. Another approach to using the cluster size data would be to stratify the counties on size and take a stratified random sample. The NHIS and NHANES samples use this strategy, with the largest clusters having $\pi_i = 1$.

Figure 3.5 shows simulation code and results for a stratified design in the US election data. The design samples the largest five counties (based in or around Los Angeles, Chicago, Houston, Phoenix, and Detroit) with certainty, then takes 5 of the next 95, 15 of the next 900, and 15 of the remaining 3600. The function one.sim() does one simulation, the call to replicate() repeats this 10,000 times, taking a few minutes to run. The results, stored in many.sims, are a matrix with six rows and 10,000 columns.

The first call to apply() computes the mean of each of the first three rows: the estimated totals. These are close to the true values shown in Figure 3.4. The second call computes the standard deviation of the estimated totals, and the third computes the mean of the estimated standard errors. These two results are similar, showing that the standard error calculation is giving accurate results. The standard errors for this design are about twice as large as for the PPS design, because the cluster size information is not being used as efficiently. It is possible to recover much of the lost efficiency, as will be shown in Section 5.1.3 and Chapter 7. It is also worth noting that the mean estimated standard error is underestimating the actual standard deviation of the estimated totals, by about 10% for Nader and about 5% for Kerry. The same techniques that improve the efficiency will also tend to alleviate the underestimation of the standard errors.

```
> one.sim<-function(){
  insample<-c(1:5,
      sample(6:100,5),
      sample(101:1000,15),
      sample(1001:nrow(sorted),15))
  strsample<-sorted[insample,]
  strsample$strat<-rep(1:4,c(5,5,15,15))
  strsample$fpc<-rep(c(5,95,900,nrow(sorted)-1000),c(5,5,15,15))
  strdesign<-svydesign(id=~1, strat=~strat, fpc=~fpc,
    data=strsample)
  totals<-svytotal(~Bush+Kerry+Nader,strdesign)
  c(total=coef(totals),se= SE(totals))
  }
> one.sim()
 total.Bush total.Kerry total.Nader     se.Bush     se.Kerry
44796129.00 46321231.00   365634.00  4403320.14  9376513.05
   se.Nader
   95575.52
>
> many.sims<-replicate(10000, one.sim())
> apply(many.sims[1:3,],1,mean)
 total.Bush total.Kerry total.Nader
 59624217.9  56165180.2    405214.7
> apply(many.sims[1:3,],1,sd)
 total.Bush total.Kerry total.Nader
  6925890.6   7549573.2    126281.5
> apply(many.sims[4:6,],1,mean)
  se.Bush  se.Kerry  se.Nader
6746625.6 7192727.8  115774.8
```

Figure 3.5 Simulating a stratified cluster sampling of US voting data

```
> svymean(~api00+meals+ell+enroll, dclus1, deff=TRUE)
           mean      SE     DEff
api00  644.1694  23.5422  9.3459
meals   50.5355   6.2690 10.4517
ell     27.6120   2.0193  2.6979
enroll 549.7158  45.1914  2.8229
> svymean(~api00+meals+ell+enroll, dclus2, deff=TRUE,na.rm=TRUE)
           mean      SE     DEff
api00  673.0943  31.0574  6.2833
meals   52.1547  10.8368 11.8585
ell     26.0128   5.9533  9.4751
enroll 526.2626  80.3410  6.1427
```

Figure 3.6   Design effects in two multistage samples of California schools

3.3.1   Loss of information from sampling clusters

Cluster sampling, even sampling proportional to size, give less precision per observation than sampling individuals. For a single-stage cluster sample with all clusters having the same number of individual m the design effect is

$$\text{DEff} = 1 + (m - 1)\rho$$

where ρ is the within-cluster correlation. Although ρ is likely to be very small, m may be very large. If $m = 100$ and $\rho = 0.01$, for example, the design effect will be 2. The formula is not very useful for quantitative forecasts, because ρ will not be known, but it indicates the general impact of cluster sampling.

Qualitatively similar increases in the design effect occur for more complicated multistage designs. For example, consider two multistage samples from the API population. The first is a cluster sample of all the schools in 15 school districts, and the second is the two-stage cluster sample described in section 3.2. Figure 3.6 shows the design effect for the means of four variables: the academic performance index (api00), the proportion of students receiving subsidized meals (meals), the proportion of 'English language learners' (ell), and the total enrollment (enroll). The design effects for all these variables are quite large, reflecting large differences between school districts in socioeconomic status, school size, and school performance. Large differences between clusters make cluster sampling less efficient because the clusters that are not represented in the sample will be different from those that are represented in the sample.

Figure 3.7 shows estimated proportions from the Behavioral Risk Factor Surveillance System 2007 data for two variables. The survey design object uses data stored in a SQLite database and accessed with the RSQLite package [66] as described in section D.2.2. X_FV5SRV reports whether the participant consumes five or more servings of fruits and vegetables per day (2=Yes, 1=No), based on a set of questions about particular food. X_CHOLCHK reports whether the participant has had a choles-

```
> library(RSQLite)
> brfss <- svydesign(id=~X_PSU, strata=~X_STATE, weight=~X_FINALWT,
data="brfss",  dbtype="SQLite", dbname="brfss07.db",
nest=TRUE)
> svymean(~factor(X_FV5SRV)+factor(X_CHOLCHK), brfss, deff=TRUE)
                        mean        SE   DEff
factor(X_FV5SRV)1   0.73096960 0.00153359 5.1632
factor(X_FV5SRV)2   0.23844253 0.00145234 5.0147
factor(X_FV5SRV)9   0.03058787 0.00069991 7.1323
factor(X_CHOLCHK)1 0.73870300 0.00168562 6.3550
factor(X_CHOLCHK)2 0.03230828 0.00058759 4.7676
factor(X_CHOLCHK)3 0.19989559 0.00162088 7.0918
factor(X_CHOLCHK)9 0.02909313 0.00055471 4.7029
```

Figure 3.7   Design effects in the Behavioral Risk Factor Surveillance System

terol measurement (1=within five years, 2=more than five years ago, 3=never). Only 24% eat the recommended amount of fruits and vegetables, but about 74% have had a recent cholesterol check. The design effects for these variables range from 4.7 to 7.1. The PSUs in the BRFSS data set are not the true PSUs in the design, which are blocks of 100 telephone numbers, but the design effect still shows the strong geographic variation in these variables.

3.4 REPEATED MEASUREMENTS

The Survey of Income and Program Participation (SIPP) is a panel survey, making repeated measurements on the same people over time. Each panel is followed for multiple years, with subsets of the panel participating in four-month waves of follow-up. The web site has a subset of the variables for wave 1 of the 1996 panel, which followed 36730 households with interviews every four months, starting in late 1995 or early 1996. The data is recorded in per-month form, so each interview results in four one-month records in the data set. The households were recruited in a two-stage sample. The first stage sampled 322 counties or groups of counties as PSUs; the second stage sampled households within these PSUs. The public use data does not contain PSU information; instead the file has "pseudo-strata" and "pseudo-PSUs" to be used for variance estimation.

The file sipp.db on the web site contains some of the variables measured in wave 1 of the 1996 SIPP panel. This is a SQLite database file with two tables, household for household-level analyses and wave1sub for person-level analyses. Figure 3.8 shows how to describe the design to R.

In model-based analyses of these data it would be important to describe and model the correlation within a person or household across the four months of data. In design-based analyses nothing special need be done, the clustering within household behaves

```
> library(RSQLite)
> sipp_hh <- svydesign(id=~ghlfsam, strata=~gvarstr, nest=TRUE,
  weight=~whfnwgt, data="household", dbtype="SQLite",
  dbname="~/SIPP/sipp.db")
> sipp_hh<-update(sipp_hh, month=factor(rhcalmn,
    levels=c(12,1,2,3,4,5,6),
    labels=c("Dec","Jan","Feb","Mar","Apr","May","Jun")))
> qinc <- svyby(~thtotinc, ~month, svyquantile, design=sipp_hh,
    quantiles=c(0.25,0.5,0.75,0.9,0.95),se=TRUE)
> round(qinc)
   rhcalmn 0.25   0.5 0.75   0.9 0.95 se.25 se.5 se.75 se.9 se.95
1        1 1300 2573 4459 6815 8958    17   26    31   47   134
2        2 1312 2608 4529 6837 8880    14   20    31   46    77
3        3 1354 2690 4685 7137 9346    15   19    29   43    79
4        4 1339 2600 4458 6757 8878    17   21    28   42    93
5        5 1408 2800 4911 7420 9539    18   30    40   71   121
6        6 1368 2601 4518 6998 9222    26   34    53   86   141
12      12 1302 2658 4683 7154 9460    24   39    55   96   192
> dotchart(qinc,cex=0.65)
```

Figure 3.8   Analyzing data from Wave 1 of the 1996 SIPP panel

like another stage of clustering in sampling and is incorporated automatically in the analysis. What is important, and may be challenging, is to use the correct weights for a longitudinal analysis.

In this SIPP example the weight variable whfnwgt is correct for estimating totals of household-level variables for a single month. These weights are also correct for computing differences between months, since a difference between two months is estimating a difference between two versions of the whole population. On the other hand, when estimating a single population total using all four months of observations for increased precision it would be necessary to divide the weight by four, as each household is represented four times. Fortunately, dividing all the weights by the same constant has no impact on estimation for other statistics such as medians, means, or regression coefficients, so this rescaling could be ignored. Figure 3.9 shows quantiles of monthly income estimated from this data set for December 1995 to June 1996. The quantiles appear quite stable over the seven-month period, although there is a suggestion that the higher quantiles decrease from December to January, i.e., that people with high income have higher income in December. This is plausible because of bonuses or because of income from retail businesses.

The situation is similar when combining waves of SIPP or years of NHANES. If the analysis estimates a single total it is necessary to rescale the sampling weights so that they add to one copy of the population, if the analysis estimates differences over time no such rescaling is needed. There is an important difference between an analysis of multiple years of NHANES and one of multiple waves of SIPP. Because

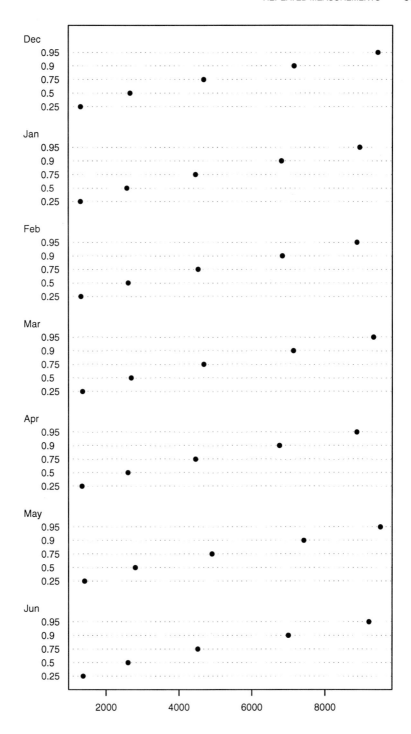

Figure 3.9 Quantiles of monthly household income for December 1995 to June 1996, estimated from wave 1 of the 1996 SIPP panel

SIPP is a panel study, different waves of data collection refer to the same people and are correlated; in NHANES different years of data collection sample different people and the sampling fraction is low enough that the data sets can be considered independent. When combining two waves of the same panel it is important to ensure that the PSUs are appropriately matched, but when combining two independent sets of data collection it is important to ensure that the PSUs are coded so that the software knows they are distinct.

The situation is more complicated when the two samples overlap partially. Details of design and estimation in this setting are beyond the scope of this book, but it is useful to mention some of the challenges. First, it is necessary to decide whether the target of estimation is a difference in population characteristics or a summary of individual-specific changes. To make the distinction clear, consider estimating differences in the mean age from two surveys conducted 10 years apart. The population as a whole will have fairly similar mean age at these two time points, for example, the US population mean age increased by about one year from 1990 to 2000. Any people present in both samples, however, will be 10 years older in the later sample than the earlier sample. The mean within-person difference is 10 years, but the difference in population means is only one year. Secondly, even if the goal is to estimate differences in population means, the individuals who are measured in both samples are more informative about the change than those measured in only sample, so that simply using sampling weights may be quite inefficient. Details for some design and estimation problems with two overlapping samples can be found in Chapter 9 of [151] and the references therein.

EXERCISES

3.1   The web site has data files demo_c.xpt of demographic data and bpx_c.xpt of blood pressure data from NHANES 2003–2004. Code to load and merge these data sets is in Appendix B, in Figure B.1.

> **a)** Construct a survey design object with these data. The sampling weights are WTMEC2YR for analysis of the clinical exam, the primary sampling units are identified by SDMVPSU and the first-stage strata by SDMVSTRA.
> **b)** Estimate the proportion of the US population with systolic blood pressure over 140 mmHg, with diastolic blood pressure over 90 mmHg, and with both.
> **c)** Estimate the design effects for these proportions.
> **d)** How do these proportions vary by age (RIDAGEYR) and gender (RIAGENDR)?

3.2   Repeat the sampling and estimation in Figure 3.4 1000 times [hint: use replicate()].

> **a)** Check that the mean of the estimated totals is close to the true population total.
> **b)** Compute the mean of the estimated standard errors and compare it to the true simulation standard error, that is, the standard deviation of the estimated totals.

c) Estimate 95% confidence intervals for the population totals and compute the proportion of intervals that contain the true population value.

3.3 The National Longitudinal Study of Youth is documented at `http://www.nlsinfo.org/nlsy97/nlsdocs/nlsy97/maintoc.html`.

    a) What are the stages of sampling and the sampling units at each stage?

    b) What would the strata and PSUs be for the single-stage approximation to the design?

3.4 The British Household Panel Survey is documented at `http://www.iser.essex.ac.uk/survey/bhps`

    a) What are the stages of sampling, the strata, and the sampling units at each stage?

    b) What would be strata and PSUs be for the single-stage approximation to the design?

3.5 Statistics New Zealand lists survey topics at `http://www.stats.govt.nz/datasets/a-z-list.htm`. Find the Household Labour Force Survey.

    a) What are the stages of sampling and the sampling units at each stage?

    b) What would be strata and PSUs be for the single-stage approximation to the design?

3.6 This exercise uses the Washington State Crime data for 2004 as the population. The data consist of crime rates and population size for the police districts (in cities/towns) and sheriffs' offices (in unincorporated areas), grouped by county.

    a) Take a simple random sample of 10 counties from the state and use all the data from the sampled counties. Estimate the total number of murders and of burglaries in the state.

    b) Stratify the sampling so that King County is sampled with 100% probability together with a simple random sample of five other counties. Estimate the total number of murders and of burglaries in the state and compare to the previous estimates.

    c) Take simple random samples of five police districts from King County and five counties from the rest of the state. Estimate the total number of murders and of burglaries in the state and compare to the previous estimates.

    d) ⋆ Take a PPS sample of 10 police districts with probability proportional to population. Estimate the total number of murders and of burglaries in the state and compare to the previous estimates.

    e) Take a simple random sample of counties and include all the police districts. Estimate the total number of murders and of burglaries in the state and compare to the previous estimates.

3.7 Use the household data from the 1996 SIPP panel (see Figure 3.8) to estimate the 25th, 50th, 75th, 90th, and 95th percentiles of income for households of different sizes (ehhnumpp) averaged over the four months. You will want to recode the large values of `ehhnumpp` to a single category (`update()`, `pmin()`). Describe the patterns you see.

3.8   In the data from the 1996 SIPP panel (see Figure 3.8),

    a) What proportion of households received any "means-tested cash benefits" (`thtrninc`)? For those households who did receive benefits, what (mean) proportion of their total income came from these benefits?

    b) What proportion of households paid rent (`tmthrnt`)? What were the mean and the 75th and 95th percentiles of the proportion of monthly income paid in rent? What proportion paid more than one-third of their income in rent?

3.9   ⋆ By the time you read this, the **survey** package is likely to provide some approximations for PPS designs that use the finite population size information. Repeat exercise 3.2 using these.

3.10   ⋆ Repeat exercise 3.2, computing the Hartley–Rao approximation and the full Horvitz–Thompson estimate using the code and joint-probability data on the web site. Compare the standard errors from these two approximations to the standard errors from the single-stage with-replacement approximation and to the true simulation standard error.

3.11   Since 1999, NHANES has been conducting surveys continuously with data released on a two-year cycle. Each data set includes a weight variable for analyzing the two-year data; the weights add to the size of the US adult, civilian, non-institutionalized population.

    a) What weight would be appropriate for estimating the number of diabetics in the population combining data from two two-year data sets?

    b) What weight would be appropriate if three two-year data sets were used?

    c) What weights would be appropriate when estimating changes, comparing the combined 1999–2000 and 2001–2002 data with the combined 2003–2004 and 2005–2006 data?

    d) How would the answers differ if the goal was to estimate a population proportion rather than a total?

CHAPTER 4

GRAPHICS

In which we see a lot just by looking.

4.1 WHY IS SURVEY DATA DIFFERENT?

Historically, survey statistics has made relatively little use of graphics. This is only partly due to the genuine difficulties in visualizing large data sets with unequal sampling weights. Other contributing factors are the use of batch-mode computing on large, centralized computers rather than interactive data analysis on individual workstations, and the traditional focus on pre-specified population summaries rather than exploratory analysis.

In social sciences and health sciences there is increasing interest in regression analysis of large data sets that have often been collected in complex survey designs, for the reasons explained in Chapters 2 and 3. This will result in the same needs for data visualization that apply in traditional cohort studies.

In model-based statistics the uses of graphics can be divided into three categories: data exploration, model criticism, and communicating results. The first and last

Complex Surveys: A Guide to Analysis Using R. By Thomas Lumley
Copyright © 2010 John Wiley & Sons, Inc.

categories carry over directly into survey inference. The second category, model criticism, appears at first glance to be less useful since survey methods require no model assumptions for validity. Looking more carefully at graphical model diagnostics, however, reveals that they are less often concerned with technical validity and more often with whether the regression coefficients are meaningful summaries of the data. This question of whether a population estimate is a useful summary is still present in design-based inference, and graphical diagnostics are thus valuable.

One area in which there is a long history of visualization for survey data is cartography. There are important practical and design challenges in finding the necessary mapping data and in drawing useful maps. There are also statistical challenges in conveying uncertainty and reducing the visual bias from varying population densities. Section 4.6 gives some examples of maps based on complex surveys, but many of the more complex issues of cartographic design and small-area spatial estimation are beyond the scope of this book. Many of these issues can be addressed in R, and some references are given.

The principal difficulty in designing graphics for complex survey data is representing the sampling weights. The graphs in this chapter use three strategies

1. Base the graph on an estimated population distribution.

2. Explicitly indicate weights on the graph.

3. Draw a simple random sample from the estimated population distribution and graph this sample instead.

A secondary problem in visualizing survey data is that data sets are often large. If a graph includes a separate plotting symbol for each point in the sample, the symbols will tend to overlap and hide each other. Graphs with thousands of plotting symbols also tend to be slow to draw and to result in large graphics files. Figure 3.3 shows the overplotting problem.

It is often difficult to determine who first used a visualization technique, and which of the later uses were independent inventions. Korn and Graubard's book [83] and 1998 paper [82] describe a number of useful graphical techniques, and the evolutionary history for some of them goes back to at least Chambers et al. [33]. Modern methods for large data sets, not specifically complex surveys, are described by Unwin et al. [177]. For a general reference on R graphics see Murrell [111] or the R Graphics Gallery at http://addictedtor.free.fr/graphiques/.

4.2 PLOTTING A TABLE

Many analyses of survey data involve tables, either tables of summary statistics as created by svyby() or contingency tables. Visualization techniques for these tables are essentially the same as if the table had been created from a simple random sample. This section describes some of the facilities that have been implemented in the survey package. Other ideas for visualizing categorical data can be found in Friendly [50].

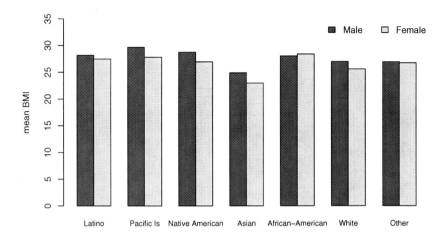

Figure 4.1 Mean BMI by race/ethnicity and gender, from CHIS

Many of these are implemented in the vcd package [108] and could easily be adapted to estimates from complex samples.

Interactive and dynamic graphics are especially valuable for displaying multivariate summaries of national or regional data. The website has links to examples of these at the UK Office of National Statistics, the SMARTCentre at the University of Durham, the Digital Government Quality Graphics project at Pennsylvania State University, and the "GapMinder" website.

Bar charts. Figures 4.1, 4.2, and 4.4 show the BMI subgroup means from CHIS 2005 that were computed with svyby() in Figure 2.11. The objects produced by most survey estimation functions have methods for the barplot() function, and so barplot(bys) will produce a barplot. The plot in Figure 4.1 has been enhanced by supplying a legend and more readable x-axis labels; the plot in Figure 4.2 is exactly as produced by dotchart(bys). Because the bars in a barplot are tied to $y = 0$ and the scales in a dotplot are not, dotplots allow the informative part of the y-axis range to be clearly seen even if it does not include zero.

Forest plots. Figure 4.4 shows the uncertainties as well as the means, in a *forest plot* (Lewis and Clarke [94]). The estimated means are depicted by squares, with the area of the square proportional to the precision (or effective sample size) for the estimate, so that more precise estimates have more visual weight. The lines extend $\sqrt{2}$ standard errors in each direction, so that overlapping confidence intervals correspond roughly to a non-significant difference at a 0.05 level. By removing the need to include zero on the range of displayed values, the forest plot expands the scale and makes differences easier to see.

The forestplot() function is in the rmeta package [98]. The code to produce this forest plot is in Figure 4.3

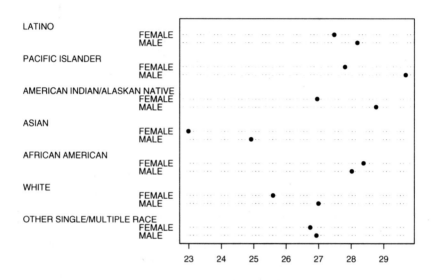

Figure 4.2 Mean BMI by race/ethnicity and gender, from CHIS

```
forestplot(ll, coef(bys), coef(bys)-sqrt(2)*SE(bys),
  coef(bys)+sqrt(2)*SE(bys), align=c("l","l"), zero=25,
  graphwidth=unit(3,"inches"), xticks=20+2*(1:5))
```

Figure 4.3 Code for forest plots of mean BMI, from CHIS

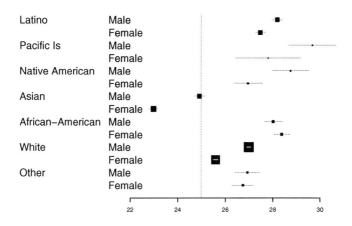

Figure 4.4   Mean BMI$\pm\sqrt{2}$ standard errors, by race/ethnicity and gender, from CHIS. The vertical line indicate the division between normal and overweight.

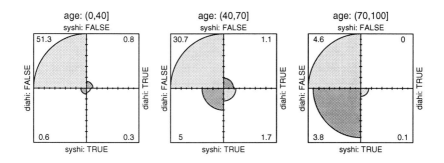

Figure 4.5   Systolic and diastolic hypertension by age group, from NHANES 2003–2004

Fourfold plots.   Figure 4.5 is a fourfold plot (Friendly [49]), a summary of a set of 2×2 tables. In this example the tables were created with svytable() and show the proportions with elevated systolic (>140 mmHg) or diastolic (>90 mmHg) blood pressure, by age group. The quarters within each table have area proportional to the population counts, with the largest cell in each table standardized to fill its box. The numbers in each cell give the percentages of the total population. Code for loading the data is in Appendix B, in Figure B.1, and the survey design was set up in exercise 3.1.

The proportion with hypertension clearly increases with age. More interestingly, the form of hypertension also changes. In the middle age group there is a substantial fraction with elevated diastolic blood pressure, but the oldest group overwhelmingly has isolated systolic hypertension, that is, high systolic pressure with normal or low diastolic pressure. We can compute the kappa measure of agreement between systolic and diastolic hypertension for different ages

```
> svykappa(~hisys+hidia,
    subset(d, RIDAGEYR<50 &!is.na(hisys) & !is.na(hidia)))
        nlcon    SE
kappa 0.41659 0.0379
> svykappa(~hisys+hidia,
    subset(d, RIDAGEYR>50 &!is.na(hisys) & !is.na(hidia)))
        nlcon    SE
kappa 0.11824 0.0226
```

and see that agreement is moderately good for people under 50 but poor for people over 50.

4.3   ONE CONTINUOUS VARIABLE

4.3.1   Graphs based on the distribution function

The most straightforward graphical displays for a single continuous variable are box-plots, cumulative distribution function plots, and survival curves. These graphs are already all based on non-parametric estimates of population quantities: the emprirical CDF is the nonparametric maximum likelihood estimator of the population CDF, the Kaplan–Meier survival curve estimates the population survival curve, and the boxplot is based on quartiles that estimate population quartiles. Adapting these estimators to complex samples simply involves inserting sampling weights in the appropriate places. Computing standard error estimates can be more difficult (e.g., Figure 2.7) but is often not necessary for graphics.

The cumulative distribution function is computed by svycdf() and the survival curve for survival data is computed by svykm(). Both functions take a formula as their first argument, with the left-hand side of the formula giving the variable to be plotted and the right-hand side giving grouping variables, and both return an object that has a plot() method for drawing a graph and a lines() method for adding the curve to an existing plot. The analogous functions for unweighted estimates are ecdf() and survfit().

Figure 4.6 is based on the stratified sample of schools from the API population defined in Figure 2.5 and shows the population CDF, the unweighted sample CDF, and the estimated CDF using sampling weights. Because the sampling design over-sampled high schools, which tend to be larger, the unweighted sample CDF shows a higher proportion of large schools than the population. The correctly weighted CDF estimate is very close to the population CDF. The code to create this figure is in Figure 4.7. The call to svycdf() estimates three CDFs, the correct one can be specified

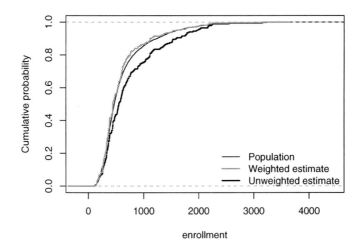

Figure 4.6   Cumulative distribution of California school size: population, weighted estimate, unweighted estimate.

by position, as in the code, or by name, e.g., cdf.est[["enroll"]]. The call to par() modifies the margins around the figure, reducing the white space at the top. The do.points argument to plot() and lines() specifies whether the individual data points should be plotted on top of the CDF line. The legend() function draws a legend, with the first argument indicating where it should be placed.

```
cdf.est<-svycdf(~enroll+api00+api99, dstrat)
cdf.pop<-ecdf(apipop$enroll)
cdf.samp<-ecdf(apistrat$enroll)
par(mar=c(4.1,4.1,1.1,2.1))
plot(cdf.pop,do.points=FALSE, xlab="enrollment",
    ylab="Cumulative probability",main="",lwd=1)
lines(cdf.samp, do.points=FALSE,lwd=2)
lines(cdf.est[[1]],lwd=2,col.vert="grey60",col.hor="grey60",
    do.points=FALSE)
legend("bottomright",lwd=c(1,2,2),bty="n",
    col=c("black","grey60","black"),
    legend=c("Population","Weighted estimate",
    "Unweighted estimate"))
```

Figure 4.7   Code for cumulative distribution function of California school size

Figure 4.8 shows the estimated survival curves for time to relapse in Wilms' tumor, using data from a two-phase case–cohort subsample of the National Wilms' Tumour

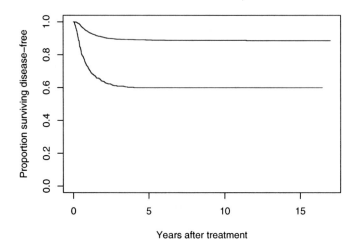

Figure 4.8   Disease-free survival in Wilms' tumor, for favorable and unfavorable histology

Study population, which is described in more detail in section 8.4.3. The code to estimate the survival curves is

```
scurves <- svykm(Surv(edrel/365.25,rel)~histol, dcchs),
```

where dcchs is name of the survey design object and the Surv() function, from the survival package wraps up the time to event and the event indicator into a single object. The graph is then drawn with plot(scurves).

The survival curves show that relapse tends to happen within about three years, when it happens at all, and that prognosis is good for the 88% of children whose tumors have favorable histological classification. In this example the use of sampling weights makes a very large difference to the estimate: the case-cohort sample includes all relapses, and so the incorrect, unweighted estimates of relapse rates are much higher than the weighted estimates.

Boxplots.   Traditional boxplots are based on quartiles of the data: the box shows the median and first and third quartiles, the whiskers extend out to the last observation within 1.5 interquartile ranges of the box ends, and all points beyond the whiskers are shown individually. svyboxplot() approximates this for a large, weighted sample. The box shows the estimated median and first and third quartiles, but the whiskers do not necessarily end at an observed point: they extend by 1.5 interquartile ranges or to the most extreme point, whichever is shorter. Only the most extreme point in each direction is shown explicitly, if it is outside the range of the whiskers.

Figure 4.9 uses the NHANES 2003–2004 demographic and blood pressure data that are loaded in section 1.4.2 and merged in section B.2.1. The survey design object uses the single-stage with-replacement approximation to the multistage NHANES design. The agegp variable is created by categorizing the reported age with cut(),

```
nhanes<-svydesign(id=~SDMVPSU, strat=~SDMVSTRA,
    weights=~WTMEC2YR, data=both, nest=TRUE)
nhanes<-update(nhanes,
    agegp=cut(RIDAGEYR,c(0,20,30,40,50,60,80,Inf), right=FALSE))
svyboxplot(BPXSAR~agegp, subset(nhanes, BPXSAR>0), col="gray80",
    varwidth=TRUE, ylab="Systolic BP (mmHg)",xlab="Age")
```

Figure 4.9   Drawing boxplots of systolic blood pressure by age

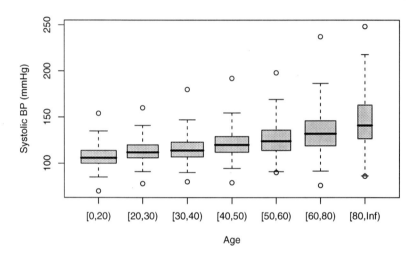

Figure 4.10   Systolic blood pressure by age in NHANES 2003–2004

and the boxplot uses only non-zero observations of blood pressure. Zero is a permitted value but implies a failure to hear the Korotkoff sounds that mark arterial flow changes and are used to define systolic and diastolic pressure. The boxplot in Figure 4.10 shows an increase in both typical level and variability of systolic blood pressure with age. The varying widths of the boxes indicate the varying population sizes for each group.

4.3.2   Graphs based on the density

Histograms and kernel density estimators in unweighted samples are designed to depict the probability density function that the data came from. Technically, when sampling from a finite population there is no smooth population density to estimate, but this is not a problem in practice. If we had complete population data we would still want to smooth or bin the data for display, and this is what sampling-weighted versions of the histogram and kernel density do.

The height of a bar in a histogram is the proportion of data in each bin divided by the width of the bin. Using sampling weights this is replaced by the estimated

```
svyhist(~BPXSAR, subset(nhanes,RIDAGEYR>20 & BPXSAR>0),main="",
    col="grey80", xlab="Systolic BP (mmHg)")
lines(svysmooth(~BPXSAR, bandwidth=5,
    subset(nhanes,RIDAGEYR>20 & BPXSAR>0)),lwd=2)
```

Figure 4.11   Drawing histograms and smooth density estimates of blood pressure.

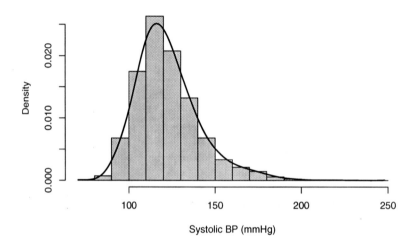

Figure 4.12   Distribution of systolic blood pressure in adults (NHANES 2003–2004)

population proportion of data in the bin divided by the width of the bin. The function svyhist() chooses the number of bins automatically, using the *ad hoc* but successful rule of choosing the same bins as for an unweighted histogram of the sample. The effectiveness of histograms is reduced by the discontinuities at the edges of the bins, and smooth density estimators can be more useful. svysmooth() fits a weighted density estimate using the local linear smoother from the KernSmooth package [182], as described by Wand and Jones [183].

Figure 4.11 shows code for a histogram and a density estimate for systolic blood pressure in adults, using the data from NHANES 2003–2004. Both svyhist() and svysmooth() take a one-sided formula as an argument to specify the variable, and a survey design object. svysmooth() also requires a bandwidth argument, analogous to the bin width for a histogram. Figure 4.12 shows the output from the code. The distribution appears to be better estimated by the smooth density, but the histogram does have the advantage that clinically important values such as 140mmHg are depicted exactly, as breaks between bins.

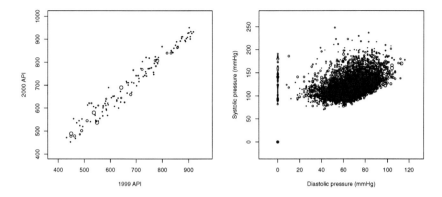

Figure 4.13   Representing sampling weights by glyph size: change in Academic Performance Index from 1999 to 2000, and relationship between systolic and diastolic blood pressure in NHANES 2003–2004.

4.4   TWO CONTINUOUS VARIABLES

The challenges of data visualization under complex sampling are clearest for scatter-plots. A scatterplot consists of a single *glyph*, or plotting symbol, for each plotted point, so it is necessary to find some way to represent the sampling weights in that glyph, and to mitigate the problems of overplotting with large numbers of points. As far as I know, there has been no empirical evaluation of any of the techniques in this section in the context of complex samples, but they all appear useful in examples. My personal preference is bubble plots for small data sets, transparent scatterplots for large data sets where color is needed, and hexagonal binning for large data sets where color is not needed.

4.4.1   Scatterplots

A direct way to represent the sampling weights is with the size of the plotted glyph, leading to "bubble plots" such as Figure 4.13. Each point is plotted as a circle with area proportional to the sampling weight. The left-hand panel shows a graph of 1999 vs 2000 Academic Performance Index based on the two-stage cluster sample described in section 3.2. The right-hand panel shows systolic and diastolic blood pressure in the data from NHANES 2003–2004 used in section 1.4.

As Figure 4.13 illustrates, bubble plots can be useful for small data sets. In the left panel, it is useful to know that the outliers above the main trend do not have large sampling weights. In the right panel, however, the individual bubbles are not visible and the bubble plot does not give a clear picture of the trend.

An alternative is to vary the intensity of shading of the points to indicate sampling weight. Figure 4.14 shows the same two data sets drawn with shading proportional to sampling weight. This partial transparency allows superimposed points to be seen, showing where the data are more or less dense. In contrast to the bubble

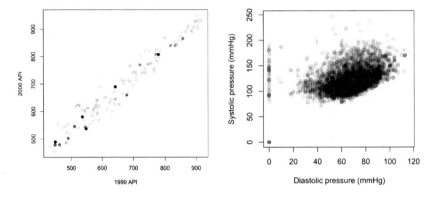

Figure 4.14 Representing sampling weights by shading: change in Academic Performance Index from 1999 to 2000, and relationship between systolic and diastolic blood pressure in NHANES 2003–2004

plot, the transparent scatterplot is more useful with large data sets. In the small sample in the left-hand panel the points with large sampling weight stand out as individuals, and the points with small weight are nearly invisible. In the right-hand panel, transparency reveals more structure in what was previously a solid blob. As Figure 4.14 unfortunately shows, transparency does not render as clearly on paper as on a computer screen.

Both bubble plots and transparent scatterplots are produced by svyplot(), which uses a model formula to specify graphs in the same way as the standard plot(), with the data coming from a survey design object rather than a data frame. The plotting style is specified by the style argument. Code for the two NHANES panels is in Figure 4.15.

```
svyplot(BPXSAR~BPXDAR,design=dhanes, style="bubble",
  xlab="Diastolic pressure (mmHg)",
  ylab="Systolic pressure (mmHg)")
svyplot(BPXSAR~BPXDAR,design=dhanes, style="transparent",
  pch=19,alpha=c(0,0.5), xlab="Diastolic pressure (mmHg)",
  ylab="Systolic pressure (mmHg)")
svyplot(BPXSAR~BPXDAR,design=dhanes, style="subsample",
   xlab="Diastolic pressure (mmHg)",
   ylab="Systolic pressure (mmHg)", sample.size=1000)
svyplot(BPXSAR~BPXDAR,dhanes,style="hex",legend=0,
   xlab="Diastolic pressure",ylab="Systolic pressure")
```

Figure 4.15 Scatterplots of systolic and diastolic blood pressure from NHANES 2003–2004

In the second line of code, which draws the transparent plot, pch=19 requests a filled circle as the plotting symbol and the alpha argument give the range of opacity

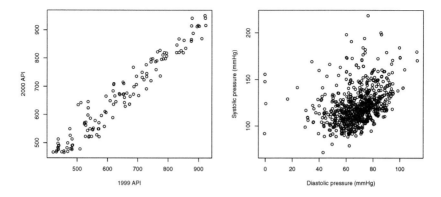

Figure 4.16   Resampling to create an unweighted scatterplot: change in Academic Performance from 1999 to 2000, and relationship between systolic and diastolic blood pressure in NHANES 2003–2004

to be used: 0 for a sampling weight of zero and 0.5 for the largest sampling weight. Bubbleplots and transparent scatterplots can be augmented by plotting the points in different colors to indicate a third, categorical variable. In R this is specified with the col argument to svyplot().

A third approach to constructing a scatterplot that is not distorted by complex sampling is to convert the data into a simple random sample by *inverse sampling*. An independent sample of size m with replacement from the data set, with probability proportional to the sampling weight, is approximately a simple random sample from the population. Points that are under-represented in the sample will have higher sampling weight and so will be subsampled with higher probability. The approximation to a simple random sample becomes better as n increases and also as m/n decreases. It is reasonable to hope that a scatterplot of the subsample of size m will be easier to interpret, because it is effectively a familiar scatterplot of a simple random sample. If the variables are continuous, a better approximation to the population distribution would result from smoothing the sample, and this can be implemented by adding random error to the m sampled values. It is always advisable to repeat a subsampled scatterplot a few times to ensure that interesting visual features are not just chance findings with one particular subsample. This subsampling approach was described by Korn and Graubard [83] (without the addition of noise for smoothing).

Figure 4.16 shows plots of the same data sets as Figures 4.14 and 4.13, using resampling from the data set. The left-hand panel has $m = n$ and added noise of 25, which is the median year-to-year change in the Academic Performance Index, and so is a reasonable "measurement error" value. The right-hand panel is a 'thinned' plot, based on a subsampled with $m = 1000$ that is much smaller than the full sample. It has just enough noise to make overlapping points detectable. Code for the right-hand panel is in Figure 4.15. The style argument to svyplot() requests a subsampled plot and sample.size specifies the subsample size m.

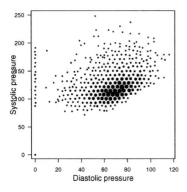 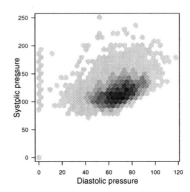

Figure 4.17 Hexagonal binned scatterplot of systolic and diastolic blood pressure from NHANES 2003–2004. The area or shading intensity is proportional to the total sampling weight in the hex.

4.4.2 Aggregation and smoothing

Plotting a symbol for each point in the data set leads to large graphics files. The transparent scatterplot of blood pressure in NHANES 2003–2004 gives a 5.5 Mb file even with only 9800 points. Screen display and printing of very large files is slow, and can even crash printer drivers or viewing software. The problems of large file size can be removed by aggregating the data or by replacing the scatterplot with a series of smooth curves giving trends in means or quantiles. These approaches scale efficiently to data sets of millions of points.

Hexagonal binning involves dividing the plotting surface into a grid of hexagons and combining all the points that fall in a grid cell into a single plotted hexagon whose shading or size indicates the number of points in the bin (Carr et al. [28]). The hexbin package [29] provides hexagonal binning and is used by the "hex" and "grayhex" styles of svyplot(). In the complex-sampling context the number of points in a hexagonal bin is replaced by the sum of the weights for points in the bin, i.e., the estimated number of individuals in the population that fall in the bin.

Figure 4.17 shows the scatterplot of blood pressures from NHANES 2003–2004 using hexagonal binning. The left-hand panel plots hexagons with area proportional to the sum of sampling weights in the bin. The right-hand panel plots hexagons of fixed size, with the shading intensity proportional to the sum of sampling weights. Both plots show the structure of the large blob of points well. In contrast to the transparent plot, individual outlying points are still visible. The graphics files for Figure 4.17 are 30–35kb in size, much smaller than those for the transparent scatterplot.

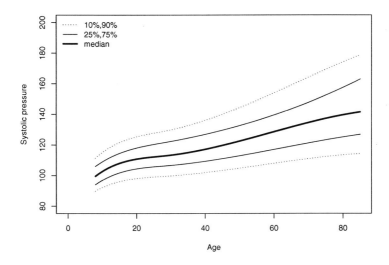

Figure 4.18   Quantiles of systolic blood pressure by age, using data from NHANES 2003–2004

4.4.3 Scatterplot smoothers

The boxplot in Figure 4.10 shows how the quartiles of systolic blood pressure vary with age in the NHANES 2003–2004 sample, but only for discrete categories of age. Figure 4.18 shows estimates of the 10th, 25th, 50th, 75th, and 90th percentiles of systolic blood pressure as smooth functions of age. As with the boxplots it is clear that blood pressure is increasing with age and that the distribution is more spread out at higher ages, presumably indicating different rates of increase for different individuals. Because the x-axis is numeric in this plot, in contrast to the discrete categories in Figure 4.10, it is easier to see the trend and to be confident that it is not being distorted by the way ages are categorized. The relationship between the smooth curves in Figure 4.18 and the boxplots in Figure 4.10 is analogous to the relationship between the histogram and smooth density estimate in Figure 4.12.

The curves are estimated by quantile regression (Koenker [79]), using the quantreg package [80]. They are similar in concept to the kernel estimators of conditional per-centiles described by Korn and Graubard [83], but are quite different in computational details. The code is given in Figure 4.19. Each call to svysmooth() estimates one quantile curve, specified by the quantile argument. The flexibility of the curve is controlled by the df argument, which is the number of degrees of freedom for a natural cubic spline curve. The calls to svysmooth() differ from those for density estimation in Figure 4.11 in having method="quantreg" to request quantile regres-sion and in having a two-sided model formula, with the left-hand side specifying the y-axis variable.

In addition to densities and smooth quantile curves, svysmooth() can also es-timate smooth mean curves. These are requested with method="locpoly" and with the smoothness of the curve specified by the bandwidth argument (as in Fig-

```
adults<-subset(dhanes, !is.na(BPXSAR))

a25<-svysmooth(BPXSAR~RIDAGEYR,method="quantreg",design=adults,
  quantile=0.25,df=4)
a50<-svysmooth(BPXSAR~RIDAGEYR,method="quantreg",design=adults,
  quantile=0.50,df=4)
a75<-svysmooth(BPXSAR~RIDAGEYR,method="quantreg",design=adults,
  quantile=0.75,df=4)
a10<-svysmooth(BPXSAR~RIDAGEYR,method="quantreg",design=adults,
  quantile=0.10,df=4)
a90<-svysmooth(BPXSAR~RIDAGEYR,method="quantreg",design=adults,
  quantile=0.90,df=4)

plot(BPXSAR~RIDAGEYR,data=both,type="n",ylim=c(80,200),
  xlab="Age",ylab="Systolic pressure")

lines(a50,lwd=3)
lines(a25,lwd=1)
lines(a75,lwd=1)
lines(a10,lty=3)
lines(a90,lty=3)

legend("topleft",legend=c("10%,90%","25%,75%","median"),
  lwd=c(1,1,3),lty=c(3,1,1),bty="n")
```

Figure 4.19 Smooth quantile functions for systolic blood pressure by age

ure 4.11) rather than the df argument. Examples of smooth mean curves will be given in Chapter 5.

For both smooth quantile and smooth mean curves the output can be quite sensitive to the amount of smoothing, as explored in exercise 4.9. For independently sampled observations there are automated techniques to choose the amount of smoothing that gives the most accurate fit to an underlying true curve, but these do not seem to have been adapted to complex samples. When smooth curves are being used for exploratory or descriptive purposes, as in the examples in this book, the "Goldilocks" approach is often adequate. That is, fit the curves with different amounts of smoothness to find examples that are "too rough" and "too smooth", and hope that an intermediate amount of smoothness will be "just right."

4.5 CONDITIONING PLOTS

A third variable can be introduced into scatterplots (and other graphs) by conditioning or 'faceting' the plot. This involves creating an array of scatterplots with the same x-axis and y-axis scales with each scatterplot showing a subset of the data based on the third variable. The conditioning variable can be discrete, in which case each plot represents a single value, or continuous, so that each plot represents a range of values. When the conditioning variable is continuous it turns out that using overlapping ranges, rather than disjoint ranges, gives more useful graphs. The use of conditioning was developed in detail by Cleveland and coworkers in the Trellis display system [4, 35], and implemented for R by Sarkar [147, 148].

Transparent scatterplots and hexagonally binned scatterplots can be produced by svycoplot(). Figure 4.20 shows hexagonally binned scatterplots conditioned on age, using the data from NHANES 2003–2004. The two plots differ in how the hexagon sizes are scaled: in the upper plot they are scaled separately for each panel, in the lower plot the scales are the same across all panels. The shaded strip above each panel shows the range of ages in that panel. Each panel has the same number of observations, and there is a 50% overlap between panels, so that most points appear twice. The overlap smooths the relationship between panels and makes it easier to see trends.

The code for creating the upper plot is

```
svycoplot(BPXSAR~BPXDAR|equal.count(RIDAGEMN), style="hex",
  design=subset(dhanes,BPXDAR>0), xbins=20,
  strip=strip.custom(var.name="AGE"),
  xlab="Diastolic pressure",ylab="Systolic pressure")
```

The conditioning variable or variables are indicated by the | symbol in the formula. In this example the conditioning variable is a *shingle*, a set of overlapping ranges, produced by the equal.count() function. The xbins arguments specifies the number of hexagonal bins, and the strip argument specifies a more readable name for the conditioning variable, to be used in the strips heading each panel. The lower plot has the additional argument hexscale="absolute" to make the hexagon scales comparable between panels.

Transparent scatterplots are specified by the style="transparent" argument to svycoplot(). They produce larger graphics files that do not print as attractively on paper, but they use the limited resolution of a computer screen more efficiently and have the advantage of being able to show an additional variable by using different colors for points. For example, in a transparent version of Figure 4.20, different colors could be used to indicate gender. The resulting plot, available on the web site, shows that blood pressure is initially lower for men, but that women catch up in the last two panels.

4.6 MAPS

4.6.1 Design and estimation issues

Many large complex surveys are national in scope, and include at least some geographic information on participants, so it is natural to compute estimates for geographic areas and display them on a map. Although specialized GIS (Geographical

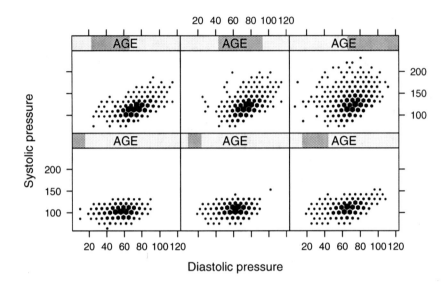

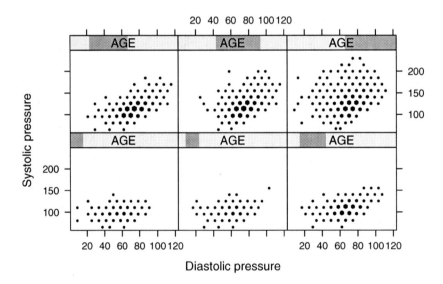

Figure 4.20 Relationship between systolic and diastolic blood pressure by age, using data from NHANES 2003–2004. In the upper plot the hexagons are scaled separately for each panel, in the lower plot the scales are the same across all panels

Information Systems) software is available, combining graphical, geographical, and database facilities, it is not always necessary and R is capable of producing useful statistical maps.

The primary operational difficulty in turning a table of regional estimates into a map is obtaining the geographical information needed to draw the map. In the United States the Census Bureau publishes the public-domain TIGER/Line database of states, counties, cities, postal codes, and streets, but in other countries such data are typically not free and may be expensive. The R packages maps [6] and mapdata [5] include US maps at the county level and world maps at the national level (but with national boundaries that are seriously outdated in areas such as Eastern Europe and Central Asia). The maptools package can read and write many popular data formats[93], making it possible to use R together with a GIS package when desirable.

A second difficulty, especially when dealing with geographical data from more than one source, is that the earth is not flat. Locations identified by latitude and longitude must be projected onto a flat page, and some distortion is inevitable. For single countries, not too close to the poles, there are many satisfactory projections, but it is still important to ensure that all data are using the same projection. When mapping small areas it becomes important that the earth is not even round: different approximations to the shape of the earth lead to differences of tens of meters in mapped locations. The R package mapproj performs a fairly wide variety of projections [106].

Figure 3.1 uses a rectangular projection for the US, which ensures that the vertical and horizontal distance scales match at the center of the map. There is definite distortion of distance at the top and bottom; a degree of longitude is nearly 10% shorter along the Canadian border than in south Florida. The US maps in this chapter use an Albers projection that accurately represents areas and has accurate horizontal scale along the 45.5N and 29.5N latitude lines.

When sample sizes are sufficiently large, computing the estimates for each geographic area is no different from any other subpopulation estimation. When sample sizes are small, however, the spatial structure provides extra opportunities to use data from adjacent regions to improve estimates. A wide range of model-based and design-based approaches have been proposed; references are Rao [129], Elliot et al. [45], and the review paper by Ghosh and Rao [52].

A further complication is that population density can be extremely variable. Regions such as census districts, zipcode regions, or counties, tend to be larger when they have smaller population, so rural areas tend to visually dominate a map. This can lead to misleading visual impressions either when rural areas are systematically different on the variables being estimated, or when the sampling uncertainty is large enough that some rural regions appear different due to chance.

A well-known example of systematic differences between rural and urban areas is the US electoral maps in which states are colored red or blue according to whether the Democratic or Republican candidate has a larger share of the vote. The "red" states have a lower average population density and so take up a much larger fraction of the map even when the voting proportions are about equal. There is no entirely satisfactory solution to this problem, but a Google search for "election 2004 maps" will show a range of attempts. Two more are on the web site for this book, based on adding color to Figure 3.1.

Color scales are an important part of map design, though not one that can be addressed well in a black-and-white book. An important distinction is between *sequential* and *diverging* scales. A diverging scale measures distance from a midpoint, a sequential scale measures distance from the endpoint(s). For example, in a US election map the most important value is the 50% at the center of the scale, and useful color schemes diverge symmetrically from this center. If the endpoints are red and blue the center point could be purple or white. A red–white–blue scale makes it easier to see the directions of small deviations from 50%, a purple scale was used by Vanderbei [180] to de-emphasize small deviations. Sequential color scales are appropriate for most of the measurements collected by the Behavioral Risk Factor Surveillance System, which is the source of the examples in this section. For the risk factors targeted by BRFSS, less is always better and only unidirectional comparisons are needed.

A useful tool for choosing color scales for maps is ColorBrewer (Harrower & Brewer [58]), which demonstrates a range of color scales designed specifically for statistical maps. These color scales are conveniently available in the RColorBrewer package [119]. A range of grays makes an effective sequential scale, although there are perceptual advantages to a color scale of uniform brightness. Diverging color schemes, on the other hand, never render well in black-and-white.

As further resources, Monmonier [110] covers the use of maps to communciate social research (the same author has written several other books on maps), Brewer [21] discusses visual design based around examples, and Elliot et al [45] examine many issues of inference relevant to mapping data from surveys.

4.6.2 Drawing maps in R

The Behavioral Risk Factor Surveillance System has developed an online mapping tool (http://apps.nccd.cdc.gov/gisbrfss/default.aspx) to display estimates for states and for metropolitan areas that have a sufficiently large sample size. The spatial data used by this application is available for download, from the BRFSS website and consists of high-resolution state outlines in the standard "shapefile" format. These have already been projected to a flat space using the Albers projection and can be plotted without further transformation.

Figure 4.22 shows code to read in the shape files, to estimate average fruit consumption in each state from the BRFSS design constructed in Figure 3.7, merge this with the shapefile data, and create a plot. The "FIPS" in the variable name is Federal Information Processing Standard 6-4, which sets out numeric codes for US counties and states [121], and the merge() call uses all=FALSE to drop estimates for Guam, Puerto Rico, and the US Virgin Islands. It is necessary to ensure that the state shape data are in order by FIPS code before merging, otherwise the relationship between shape data and variables can be destroyed. The function spplot() from the sp package [125] draws the map using the shape data from the states object and fills in

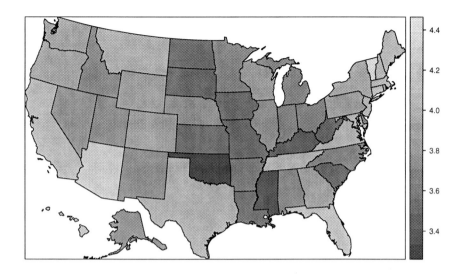

Figure 4.21 Servings of fruit per day, from BRFSS 2007

```
library(maptools)
states<-readShapePoly("brfss_state_2007_download")
states<-states[order(states$ST_FIPS),]

bys<-svyby(~X_FRTSERV,~X_STATE,svymean,
    design=subset(brfss,X_FRTSERV<99999))

states@data<-merge(states@data, bys, by.x="ST_FIPS",
    by.y="X_STATE",all=FALSE)
states$fruitday<-states$X_FRTSERV/100

spplot(states,"fruitday")
```

Figure 4.22 Drawing a map of average fruit consumption from BRFSS 2007

each state according to the variable `fruitday`, which is average number of servings of fruit per day.

Figure 4.23 creates a set of four linked maps showing health insurance coverage by age. The state map data and the BRFSS survey data are the same as in Figure 4.22, but the estimation step uses age groups ≤ 35, 35–50, 50–65, and > 65 in addition to state to specify subpopulations. The `reshape()` function reshapes the data set of estimates from 4×54 records in one column to 54 records in four columns.

The data frame is merged with the state shape data in the same way as the previous example. Finally, the call to `spplot()` specifies that the variables `age1`–`age4` should

```
brfss<-update(brfss, agegp=cut(AGE, c(0,35,50,65,Inf)))
hlth<-svyby(~I(HLTHPLAN==1), ~agegp+X_STATE, svymean,
    design=brfss)

hlthdata<-reshape(hlth[,c(1,2,4)],idvar="X_STATE",
    direction="wide",timevar="agegp")
names(hlthdata)[2:5]<-paste("age",1:4,sep="")

states@data<-merge(states,hlthdata,
    by.x="ST_FIPS",by.y="X_STATE",all=FALSE)
spplot(states,c("age1","age2","age3","age4"),
    names.attr=c("<35","35-50","50-65","65+"))
```

Figure 4.23   Drawing a map of health insurance coverage by age with BRFSS 2007

be used to color the states, resulting in a set of four graphs with the same color scale. The names.attr option specifies names for each of the maps, corresponding to the age groups. Figure 4.24 shows the result. Health insurance coverage is very high for people over 65, because Medicare provides nearly universal coverage to citizens and permanent residents in this age group. Coverage is lower and more variable at younger ages. In these younger age groups coverage appears to be particularly low in Nevada and Texas and high in Wisconsin, Minnesota, and Massachusetts. Since Nevada has a fairly small population it is important to check the standard errors. The standard error for Nevada is only 2.6%, so the low insurance coverage is not just sampling variation.

BRFSS also provides summaries of some variables for smaller areas — cities and counties, called MMSAs — through the SMART system. These smaller areas are those where at least 500 responses were received, allowing sufficient precision in estimating local totals. Figure 4.25 shows insurance coverage for states and for the MMSAs on the same map. This requires more effort to ensure that the same color scale is used for the regions as for the points. The map was constructed with the ggplot2 package (Wickham[186]), which implements Wilkinson's [187] "Grammar of Graphics". The code is on the web site for the book.

Most of the MMSAs have similar insurance coverage rates to their surrounding states. There are some fairly obvious exceptions near the Texas–Mexico border with much lower coverage rates, and some less obvious ones in inland California, central Washington, and Florida.

The version of this map on the BRFSS web site uses a color scale with five discrete steps, in which all of Texas in the lowest step, so the MMSAs on the Texas–Mexico border do not show up, but those in California and Washington are clearly visible as being in a lower step than their surrounding states. Figure 4.26 shows the data using a five-level scale than puts about the same number of states into each level, which is close to the online BRFSS map. This example shows how large the impact of

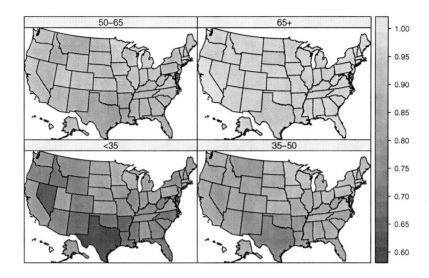

Figure 4.24   Health insurance by age, from BRFSS 2007

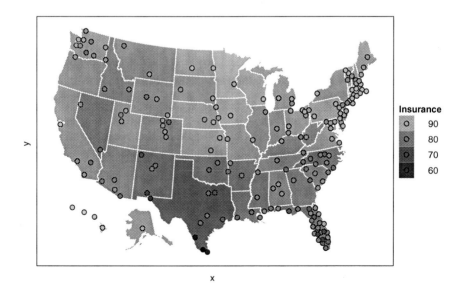

Figure 4.25   Health insurance for states and MMSAs, from BRFSS 2007

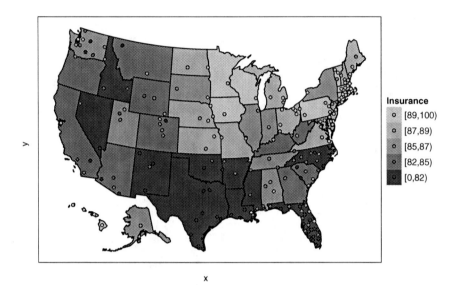

Figure 4.26   Health insurance for states and MMSAs, with a quantile-based color scale

different color scales can be. Because the low values are much lower than the rest of the country, a linear mapping of estimate to grayscale gives quite a different picture to a quantile-based mapping.

EXERCISES

4.1   Draw boxplots of body mass index by race/ethnicity and by sex using the CHIS 2005 data introduced in Chapter 2.

4.2   Using the code in Figure 3.8 draw a barplot of the quantiles of income and compare it to the dotchart in Figure 3.9. What are some advantages and disadvantages of each display?

4.3   Use svysmooth() to draw a graph showing change in systolic and diastolic blood pressure over time in the NHANES 2003–2004 data. Can you see the change to isolated systolic hypertension in old age that is shown in Figure 4.5?

4.4   With the data from the SIPP 1996 panel (see Figure 3.8) draw the cumulative distribution function, density function, a histogram, and a boxplot of total household income. Compare these graphs for their usefulness in showing the distribution of income.

4.5   With the data from the SIPP 1996 panel (see Figure 3.8) draw a graph showing amount of rent (tmthrent) and proportion of income paid in rent. You will want to exclude some outlying households that report much higher rent than income.

4.6   Using data from CHIS 2005 (see section 2.3.1) examine how body mass index varies with age as we did with blood pressure in this chapter.

4.7   The left-hand panel of Figure 3.3 shows an interesting two-lobed pattern. Can you find what makes these lobes?

4.8   Set up a survey design for the BRFSS 2007 data as in Figure 3.7. BRFSS measured annual income (income2 in categories < $10k, $10–15k, $15–20k, $20–$25k, $25–35k, $25–50k, $50–75k, > $75k) and race (orace: white, black, asian, native hawaiian/pacific island, native american, other).

   a) Draw a graph of income by race.
   b) Draw maps showing the geographical distribution of income and of racial groups.
   c) Draw a set of maps examining whether the geographical distribution of income differs by race.

4.9   Explore the impact on the graphs in Figure 4.18 of changes in the amount of smoothing, by altering the df argument to the code in Figure 4.19.

CHAPTER 5

RATIOS AND LINEAR REGRESSION

In which a line must be drawn.

Regression has two main uses in analyzing complex surveys. The first is the same as for any other type of data: to learn about relationships between variables. The second is more specialized: a regression model can give more accurate estimates of population means and totals.

All the statistical inference in this chapter is still design-based and its validity makes no assumptions about the truth of the model or the distribution of any variable. Regardless of the model, for example, confidence intervals for regression slopes will have approximately a 95% chance of including the true slope, which is defined as the value that would be obtained by fitting the same regression model to complete population data.

The validity of inference regardless of model does not stop model choice from being an important issue. Any model can estimate a summary of the population distribution, but only some models estimate useful summaries. Any model can give an approximately unbiased estimate of the population mean or total, but only some models give usefully precise estimates. The design-based view of regression

Complex Surveys: A Guide to Analysis Using R. By Thomas Lumley
Copyright © 2010 John Wiley & Sons, Inc. **83**

assists in model choice by clarifying that we are interested in the interpretation and statistical precision of the summaries resulting from the model, and that questions of the distribution of the data and the structure of relationships between variables are relevant only to the extent that they affect the precision and meaning of these summaries.

5.1   RATIO ESTIMATION

Ratios of population means or totals are important for three reasons: the ratio may be directly of interest, it may be used to estimate a population mean or total, or it may be used to construct a subpopulation estimate of a mean.

5.1.1   Estimating ratios

In the Academic Performance Index population there is a variable api.stu, counting the number of students who took the Academic Performance Index exams. The proportion of students who took the exams, across the whole state, is the population total for api.stu divided by the total number of students, i.e., the population total for the enroll variable. Given a sample from this population we might want to estimate the proportion of students who took the exams. Many quantities of economic interest are ratios, for example the national savings ratio is based on the ratio of income to expenditure, and the inflation rate is a ratio of past to current prices. These will have to be estimated from survey data.

It is important to distinguish ratios of population totals from summaries of individual-level ratios. For example, the ratio of HDL (or "good") cholesterol to LDL (or "bad") cholesterol is an important risk indicator for heart disease. This is a number characteristic of an individual, not of a population. We might be interested in how many people have an HDL:LDL ratio less than 1/4, or in the quartiles or mean of the HDL:LDL ratio over a population. There would be no reason to estimate the ratio of population totals: the population HDL cholesterol total divided by the population LDL cholesterol total is not an interesting quantity.

The function svyratio() estimates ratios of population totals. It takes two model-formula arguments, one specifying numerator variables and the other specifying denominator variables, and estimates the ratios for each pair of variables. For example, in the stratified sample of California schools created in Figure 2.5 the following code estimates the proportion of students who took the API tests:

```
> svyratio(~api.stu,~enroll, strat_design)
Ratio estimator: svyratio(~api.stu, ~enroll, dstrat)
Ratios=
          enroll
api.stu 0.8369569
SEs=
          enroll
api.stu 0.007757103
```

The true proportion based on the population data is 0.8387, so the estimate is quite accurate.

Ratio estimates are not exactly unbiased, but in large samples the bias is much smaller than the standard error, and so is pratically negligible; the bias is approximately proportional to $1/n$ and the standard error is proportional to $1/\sqrt{n}$. We will refer to estimators like this as "approximately unbiased", a term that is defined precisely in Appendix A.1.2.

5.1.2 Ratios for subpopulation estimates

If we had had individual student data rather than school data the ratio estimated in the previous section would have been a simple proportion. Proportions in subpopulations are ratios: the proportion of people over 65 who have high blood pressure is the number of people over 65 with high blood pressure divided by the number of people over 65. The totals in both the numerator and denominator must be estimated. Means in subpopulations are also ratios: the mean body mass index for African-American men in California, estimated as 28.0 with svyby() in Figure 2.11, is the total BMI for African-American men in California, divided by the number of African-American men in California. Even population means are ratios, since they are computed as population totals divided by *estimated* population size, as we saw in section 2.1.

Since a wide range of software now handles subpopulation estimates correctly in complex surveys there is no need for the data analyst to know that the computations are being done as ratio estimates, but it may help explain why special software is needed and why it is not valid just to throw away all the data outside the subpopulation. If you are happier simply knowing that the survey package does the right calculations, you can skip to the next subsection and ignore Figure 5.1.

Figure 5.1 estimates the mean BMI for African-Americans in California using data from CHIS in three ways: with svymean(), with svyratio(), and by estimating the numerator and denominator totals and using svycontrast(). The survey design object chis was defined in section 2.3.1. To use svyratio() and to compute the totals it is necessary to have an indicator variable for the subpopulation of African-American men, so the variable is_aamale is added to the design object using update(). The ratio and total estimates then make use of the fact that a logical (TRUE/FALSE) variable is converted to 1 for TRUE and 0 for FALSE when used in arithmetic expressions. The call to uses backquotes around the estimate names in the output of svytotal() because these are not legal variable names. All three approaches give the same answer, 28.019 with a standard error of 0.2663.

5.1.3 Ratio estimators of totals

The third use for ratios in analyzing complex surveys is in constructing more accurate estimates of population means or totals. Consider the Academic Performance Index data analyzed in the previous section. From a sample of 200 schools we estimated that $83.69\% \pm 0.77\%$ of students take the API tests. Suppose we know the total enrollment for all schools in California (3811472), and we want to estimate the total number who

```
> svymean(~bmi_p,
    subset(chis, srsex=="MALE" & racehpr=="AFRICAN AMERICAN"))
        mean      SE
bmi_p 28.019 0.2663
> chis<-update(chis,
    is_aamale = (srsex=="MALE" & racehpr=="AFRICAN AMERICAN"))
> svyratio(~I(is_aamale*bmi_p), ~is_aamale, chis)
Ratio estimator: svyratio(~I(is_aamale*bmi_p), ~is_aamale, chis)
Ratios=
                        is_aamale
I(is_aamale * bmi_p)   28.01857
SEs=
          [,1]
[1,] 0.266306
> totals<-svytotal(~I(bmi_p*is_aamale)+is_aamale, chis)
> totals
                        mean        SE
I(bmi_p * is_aamale) 19348927 260401.7
is_aamaleFALSE       25697039   6432.3
is_aamaleTRUE          690575   6341.9
> svycontrast(totals,
    quote('I(bmi_p * is_aamale)'/'is_aamaleTRUE'))
         nlcon    SE
contrast 28.019 0.2663
```

Figure 5.1   Subpopulation means compared to ratio estimates using data from CHIS

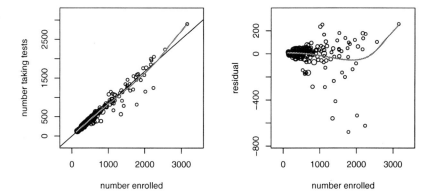

Figure 5.2   Number of students taking the API tests and number enrolled, and residuals from a ratio model

took the API tests. An obvious estimate would be $0.8369 \times 3811472 = 3189821$, applying the estimated proportion from the sample to the whole population. The standard error of this ratio estimate of the total is the standard error of the ratio multiplied by the known denominator, i.e., $0.0077 \times 3811472 = 29000$. The Horvitz–Thompson estimate of the number who took the API tests is 3086009 with a standard error of 99500, and the true population value is 3196602. Using the estimated ratio together with population information on the denominator has reduced the standard error by a factor of three. Although this computation can easily be performed by hand, there is predict() method to do the computations based on the output of svyratio()

```
> r <- svyratio(~api.stu, ~enroll, strat_design)
> predict(r, total= 3811472)
$total
          enroll
api.stu 3190038

$se
          enroll
api.stu 29565.98
```

Using the ratio in this way is motivated by the idea that the number of students taking the test will be roughly proportional to the number of students in the school, i.e., by a model

$$api.stu = \alpha \times enroll + \epsilon$$

with ϵ having approximately zero mean and with α being the ratio. Figure 5.2 shows that this model is a good approximation to the distribution of data. The left-hand panel shows api.stu and enroll, with the straight line corresponding to the model and a smooth curve estimated by svysmooth(). The right-hand panel shows the residuals

$$\text{resid} = \text{api.stu} - 0.8369 \times \text{enroll}$$

with a smooth curve fitted through them. The variability of the residuals seems to increase as the school size increases. This would be expected, since the residual cannot be larger than the total enrollment. In fact, the residual variance is roughly proportional to enroll^2. There is a suggestion that the ratio is systematically lower for schools of 1000–2000 students so that the model does not fit perfectly.

The estimate of 3189821 and standard error of 29000 for the ratio estimator of the population total did not involve any consideration of model choice or goodness of fit. The ratio estimator is approximately unbiased and the standard errors are correct regardless of whether there is a linear relationship between api.stu and enroll. On the other hand, the gain in precision from using the ratio estimator is dependent on the model fit. When a linear model with zero intercept explains most of the variation, as it does here, a ratio estimator will be much more precise than the simple Horvitz–Thompson estimator. In addition, when the residual variance increases with fitted value, as it does here, estimating the slope α by the ratio of estimated population totals will be highly efficient (the ratio estimator is optimal when the variance is proportional to the fitted mean).

If the model fits well, the residuals will have zero mean for each value of enroll and small variance. If the model fits poorly, the residuals may not have zero mean for each value of enroll, but they will still have zero mean taken over the whole population. In fact, the Horvitz–Thompson estimator of the population total of the residuals is exactly zero:

$$
\begin{aligned}
\sum_{i=1}^{n} \frac{1}{\pi_i} \text{resid}_i &= \sum_{i=1}^{n} \frac{1}{\pi_i} \text{api.stu}_i - \hat{\alpha} \times \sum_{i=1}^{n} \frac{1}{\pi_i} \text{enroll} \\
&= \sum_{i=1}^{n} \frac{1}{\pi_i} \text{api.stu}_i - \frac{\sum_{i=1}^{n} \frac{1}{\pi_i} \text{api.stu}_i}{\sum_{i=1}^{n} \frac{1}{\pi_i} \text{enroll}} \times \sum_{i=1}^{n} \frac{1}{\pi_i} \text{enroll} \\
&= 0.
\end{aligned}
$$

The use of estimators that are valid regardless of model fit and efficient when the model fits well is called *model-assisted inference*. We will refer to the model as a *working model*, a term used in similar contexts in longitudinal data analysis [191], to indicate that there is no assumption of model accuracy, that it is just a practical tool to get more precise estimates.

We have not mentioned the distribution of ϵ, because it is of little importance. Even though ϵ is far from Normal, $\hat{\alpha}$ and the estimated population total will have an approximately Normal distribution. Figure 5.3 shows the distribution of 10,000 ratio estimates of population totals for api.stu based on 10,000 stratified samples of size 200 from the API population. As we saw for estimated means in Figure 2.1, estimates based on reasonably large samples have approximate Normal distributions whatever the distribution of the data.

To illustrate that the validity of the ratio estimator does not depend on a linear relationship, we can construct a ratio estimator using a completely unsuitable de-

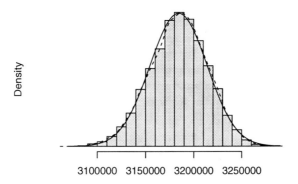

Figure 5.3   Sampling distribution of the ratio estimate of population total, based on 10,000 stratified samples of size 200 from the API population, with kernel density estimate (dotted line) and Normal density (solid line) for comparison

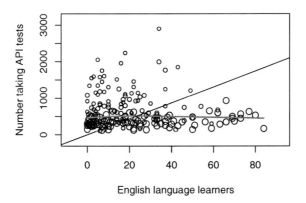

Figure 5.4   Number taking API tests and percentage of "English language learners" in California schools, with smooth fitted mean curve and ratio working model

nominator variable. Figure 5.4 shows the relationship between `api.stu` and the percentage of "English language learners" in the school. The working model of a linear relationship through the origin does not fit at all well, and the `svysmooth()` curve shows that the two variables are almost unrelated. The estimated ratio is 21.79, with a standard error of 1.60, and the population total for the denominator variable is 141685, so the estimated population total for students taking the tests is 3087479, with a standard error of 227000. The estimate is still approximately unbiased (and in fact is close to the Horvitz–Thompson estimate) but the standard error is 2.5 times higher than for the Horvitz–Thompson estimate and nearly 10 times higher than for the ratio estimator using school size as the denominator.

5.2   LINEAR REGRESSION

5.2.1   The least-squares slope as an estimated population summary

Linear regression summarizes the difference in mean of a *response* variable Y over different values of one or more *predictor* variables X. Other names for Y are the "outcome" or "dependen't" variable, and other names for X are "independent" or "carrier" variables. The working model for linear regression comes in two parts. The *systematic* part of the model describes the differences in mean

$$E[Y] = \alpha + X\beta \tag{5.1}$$

and the *random* part of the working model says that the variance of Y is constant

$$\text{var}[Y] = \sigma^2. \tag{5.2}$$

These parts of the model have different levels of importance. The systematic part of the working model defines the meaning of the coefficients: β is the average difference in Y for observations differing by 1 unit on X. The random part of the model affects how observations are weighted in estimating α and β; it does not affect the meaning of the parameters but does affect the precision of estimates.

In a model-based analysis it would be necessary to specify the random part of the model correctly to get correct standard errors, but all our standard error estimates are design-based and so are valid regardless of the model. It is worth noting that the "sandwich", or "model-robust", or "heteroskedasticity-consistent" standard errors [64, 185] sometimes used in model-based regression analysis are almost identical to the design-based standard errors we will use; the main difference being in the handling of stratification.

The parameters in the regression model will be estimated by sampling-weighted least squares. If we had complete population data we would define α and β to be values that minimized the sum of squared residuals over the population

$$RSS = \sum_{i=1}^{N} (Y_i - \alpha - X_i\beta)^2.$$

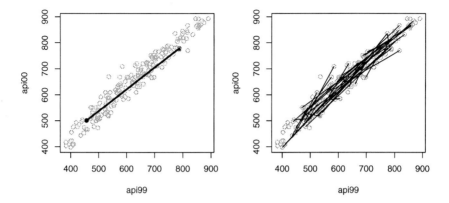

Figure 5.5   The ordinary linear regression estimator is a weighted average of all pairwise slopes

With data from a complex sample we can estimate the population sum of squared residuals using the sampling weights, as

$$\widehat{RSS} = \sum_{i=1}^{n} \frac{1}{\pi_i} (Y_i - \alpha - X_i \beta)^2,$$

and estimate α and β by the values that minimize this estimated population total.

To illustrate the point that the interpretation of β as an average difference in Y per unit difference in X makes sense independent of the fit of the model, consider a data set with just two points (x_1, y_1) and (x_2, y_2). The obvious estimate of β would be the slope of the line between the two points

$$\hat{\beta}_{1,2} = \frac{y_1 - y_2}{x_1 - x_2}.$$

For the line shown in the left-hand panel of Figure 5.5, the difference in x is 330 and the difference in y is 277, giving a pairwise slope of 0.839.

If we have n points from a simple random sample there are $n(n-1)/2$ such pairs of points, so we could compute $n(n-1)/2$ different slopes, with $\hat{\beta}_{i,j}$ being the slope of the line between (x_i, y_i) and (x_j, y_j). Figure 5.5 shows 50 such lines. These estimates can be combined by averaging them, but there is more information when $x_i - x_j$ is large, so we would prefer a weighted average. It is clear in Figure 5.5 that short lines have much more variable slopes than longer lines. If we used weights $w_{ij} = (x_i - x_j)^2$ the weighted average would be

$$\hat{\beta} = \frac{\sum_{i,j=1}^{n} w_{ij} \hat{\beta}_{ij}}{\sum_{i,j=1}^{n} w_{ij}}$$

and this is exactly the least-squares estimate of β from ordinary linear regression.

For data from a complex design the sampling weights make the calculation of $\hat{\beta}$ slightly more complicated, but the basic principle still holds that the regression coefficient estimates can be interpreted as average differences in Y per unit difference in X in a way that does not rely at all on assumption about a model or about goodness of fit.

5.2.2 Regression estimation of population totals

The ratio estimator of a population total is based on a working model that fits a straight line with zero intercept and a single predictor variable. A slight generalization of this is the *separate ratio estimator*, for a stratified design where the denominator total is known separately for each stratum, not just for the whole population. in which a ratio estimator for the population total in each stratum is computed and these estimated stratum totals are added. The separate ratio estimator corresponds to a working model

$$E[Y_i] = \beta_k \times X_i \times \{i \text{ is in stratum } k\}$$

where β_k is a ratio for stratum k. In large sample sizes the separate ratio estimator will be more accurate than the estimator based on a single ratio, because it uses more population information and because the larger number of estimated parameters allow the working model to fit the data better. In the example from section 5.1.3 the estimate and standard error for the separate ratio model are almost identical to those for a single ratio, because the proportion of students taking the API tests does not differ appreciably between elementary schools, middle schools, and high schools. Figure 5.6 shows the computations for the separate ratio estimator, and the comparison of ratios between school types.

The process of expanding the working model to incorporate more population information and fit the data better can be continued. Any regression model can be used, as long as population totals are available for all the variables in the working model. A first step is to add intercept parameters to the working model for ratio estimation, rather than forcing the intercept to be zero. This step is important because it removes the possibility that a poorly chosen set of predictors could give a much worse estimate than just using the Horvitz–Thompson estimator.

We saw that with a ratio estimator, using the percentage of "English language learners" as the denominator variable for estimating the number of students taking the API tests gave a standard error more than twice as large as for the Horvitz–Thompson estimator. The working model with an intercept added is

$$\texttt{api.stu} = \alpha + \beta \times \texttt{ell} + \epsilon.$$

Since the two variables are almost unrelated, $\hat{\beta} \approx 0$ and $\hat{\alpha}$ is almost the same as the estimated population mean of $\texttt{api.stu}$. When $\hat{\beta} = 0$ the regression estimator for the total is just $N\hat{\alpha}$, and the estimate and standard error are almost identical to those for the Horvitz–Thompson estimator. This result is true quite generally. Adding an extra variable to a regression model for estimating a population total can at most increase the standard error by a small amount. This small amount of extra variability comes

```
> sep<-svyratio(~api.stu,~enroll, strat_design,separate=TRUE)
> com<-svyratio(~api.stu, ~enroll, strat_design)
> stratum.totals<-list(E=1877350, H=1013824, M=920298)
> predict(sep, total=stratum.totals)
$total
        enroll
api.stu 3190022
$se
          enroll
api.stu 29756.44
> predict(com, total= 3811472)
$total
        enroll
api.stu 3190038
$se
          enroll
api.stu 29565.98
> svyby(~api.stu, ~stype, design=dstrat, denom=~enroll, svyratio)
  stype        V1         se
E     E 0.8518163 0.00703236
H     H 0.8105702 0.02047726
M     M 0.8356958 0.01818744
```

Figure 5.6   Estimating the number of students taking the API tests with a separate ratio for each stratum and a common ratio across strata

```
frs.des <- svydesign(id=~PSU, weights=~GROSS2, data=frs)
svymean(~HHINC, subset(frs.des,  ADULTH==1 & DEPCHLDH>0),
    deff=TRUE)
svymean(~HHINC, subset(frs.des,  DEPCHLDH>0),deff=TRUE)
svymean(~HHINC, frs.des,deff=TRUE)

frs.des <- update(frs.des,
  ctband1=CTBAND==1, ctband2=CTBAND==2,
  ctband3=CTBAND==3, ctband4=CTBAND==4,ctband5=CTBAND==5,
  ctband6=CTBAND==6,ctband7=CTBAND==7, ctband8=CTBAND==8,
  ctband9=CTBAND==9,
  tenure1=TENURE==1, tenure2=TENURE==2, tenure3=TENURE==3,
  tenure4=TENURE==4)

m <- svyglm(HHINC~ctband2+ctband3+ctband4+ctband5+ctband6
  +ctband7+ctband8+ctband9+tenure2+tenure3+tenure4,
  design=frs.des)
totals <- c(2236979, 547548, 351599, 291425, 266257, 147851,
    87767, 9190, 19670, 493237, 128189, 156348)
names(totals) <- c("(Intercept)", "ctband2", "ctband3",
    "ctband4", "ctband5", "ctband6", "ctband7", "ctband8",
    "ctband9", "tenure2", "tenure3", "tenure4")
totals <- as.data.frame(t(totals))
totincome <- predict(m, newdata=totals, total= 2236979)
svycontrast(totincome, 1/2236979)
```

Figure 5.7   Estimating mean incomes in Scotland

from the need to estimate another coeffiicent, and it becomes smaller as the sample size increases. The worst case occurs when the coefficient for the extra variable is approximately zero so that the variable is effectively ignored in estimating the total.

Example: Incomes in Scotland. A subset of the Family Resources Survey with the data for Scotland is available on the web site, via a link to the PEAS project at Napier University. This is a set of 4695 observations of households in a cluster sample. We can define a survey design and estimate the mean income for all households, households with children, and single-parent households, using the code in Figure 5.7 The mean weekly incomes are £483 (std error £10.6) for all households, £611 (std error £15.6) for all households with children, and £277 (std error £8.5) for single-parent households. The design effect decreases from 2.9 for all households to 1.01 for single-parent households, showing that over-sampling poorer households has been successful.

Population totals are available for two categorical variables, council tax band (CTBAND) and housing tenure (TENURE), i.e., whether people rent privately, rent from the government, own, or have a mortgage on a house. Figure 5.7 defines indicator variables for each category and then fits a linear model and predicts the total household income. The total can be divided by the number of households in Scotland (2236979) to obtain the mean. The reason that indicator variables had to be defined by hand, rather than using R's automatic handling of categories, is that the `predict()` function needs a single row of data to predict the total, and so needs explicit, named totals for each of the indicator variables.

The estimate from the regression model for the mean weekly househould income is £483, with a standard error of £7.5. The standard error has been considerably reduced by using the known population information. Surprisingly, the point estimate is identical. This is because the weights in the survey have already been adjusted using information about population totals. Large-scale national surveys routinely do these adjustments, which are described in the next chapter.

It is more difficult to construct a regression estimator to improve the estimates of mean weekly income for households with children and for single-parent households, as population information is not available for these subsets. In the Chapter 7 we will see how to use this information, but also see that it does not provide much extra precision.

Example: US elections.   The `elections` data set on the web site is based on county-level voting totals for the US presidential elections in 2000 and 2008. In both years these are preliminary totals, and some states are not broken down by county, so the numerical results in this example will not match exactly the official results. The same analysis can be done with high-quality data available commercially from `http://www.uselectionatlas.org/` with essentially the same conclusions.

There are 3049 regions (counties or states) in the data set, and the total vote was 65627076 for Barack Obama and 57545471 for John McCain. If we take a simple random sample of 200 regions we can define a survey design object by

```
srsdes<-svydesign(id=~1, data=elections[in.srs,], fpc=~fpc)
```

Using the Horvitz–Thompson estimator the estimated totals from this design are 77 million for Obama and 56 million for McCain, with standard errors of 29 million and 13 million, respectively.

We have data from the 2000 presidential election for all the regions, and so we can fit linear regression models and compute predictions as in Figure 5.8. The predictions from the linear regression model are 58 million for Obama and 51 million for McCain, with standard errors of 6 million and 4 million, respectively, a very large improvement in precision.

The working linear model has constant variance around the regression line, but we would expect more variability in larger regions. We can fit a working model in which the variance around the regression line is proportional to the fitted value, and so effectively proportional to number of votes in the region. The third block of code in Figure 5.8 fits these working models and predicts from them. The resulting estimates are 61 million for Obama and 54 million for McCain, with standard errors of 4.1 million and 3.7 million, a further improvement in both accuracy and standard error.

```
m.obama<-svyglm(OBAMA~BUSH+GORE, srsdes)
m.mccain<-svyglm(MCCAIN~BUSH+GORE, srsdes)

data2000<-data.frame(BUSH=sum(elections$BUSH),
        GORE=sum(elections$GORE))
predict(m.obama, total=3049, newdata=data2000)
predict(m.mccain, total=3049, newdata=data2000)

l.mccain<-svyglm(MCCAIN~BUSH+GORE, srsdes,
    family=quasi(variance="mu"))
l.obama<-svyglm(OBAMA~BUSH+GORE, srsdes,
    family=quasi(variance="mu"))

predict(l.obama, total=3049, newdata=data2000)
predict(l.mccain, total=3049, newdata=data2000)
```

Figure 5.8   Estimating US election results from a sample of counties

These improvements are not really surprising: it is well-known that "swings" from the same regions in previous elections are the best way to use early election data. This example is unrealistic because early election returns are not a random sample, but the regression model can also help in reducing bias when the data are not a random sample because of non-response. As an illustration, suppose that instead our sample consists of all 225 counties in Indiana and Virginia, two states that are among the first to report their results. These counties are not very representative of the population, and we would expect considerable bias. The estimated totals are 44 million for Obama and 41 million for McCain, with standard errors of 6 million and 4 million, underestimating both the overall total and Obama's margin. Estimates based on a linear regression model are 65 million for Obama and 49 million for McCain, with standard errors of 8 million and 3 million. Using the regression model and the known population data for the 2000 elections has considerably reduced the bias. Obama's vote is now overestimated, but only by about one estimated standard error.

Estimated population totals from a working regression model are always valid if the data are a probability sample, and will be more precise if the working model fits well. If the data are not a probability sample, perhaps because of non-response, the estimates will be less biased if the working model fits well, and may also be more precise. In either situation, using known information from the population gives better estimates of unknown information. We will see the election example again in Chapter 7, illustrating other ways of using known population data. The remainder of the current chapter will focus on fitting regression models to estimate relationships between variables.

5.2.3 Confounding and other criteria for model choice

The choice of predictor variables in a working linear regression model is important because it affects the meaning of the coefficients. Some of these choices are purely at the discretion of the analyst: do you want to compare smokers to non-smokers or compare different histories of smoking measured in pack-years? Others are forced by the structure of the scientific question: if you want to know whether coffee consumption affects blood pressure, the comparisons of blood pressure are only meaningful when they are between groups of coffee drinkers and non-drinkers who are similar on factors such as weight and age.

Useful predictor variables can be divided into three categories:

Exposure of interest: The substantive questions are about the relationship between this variable and the response. The analyst is free to choose how to summarize this relationship, whether as a difference between categories, a linear trend, or a detailed exposure–response curve. The choice will depend on sample size, the level of prior knowledge, and the use to which the conclusions will be put.

Confounding variables: These measure other, uninteresting, reasons why X and Y would be related. These must be modeled accurately in order for the uninteresting associations to be controlled. Their coefficients are usually not of interest in themselves.

Precision variables: These are unrelated to the exposure of interest and so do not affect the interpretation of its coefficient, but they explain some of the variability in Y and so reduce the standard errors. Again, their coefficients are not usually of direct interest. Precision variables are only useful if they have strong relationships with the response; reducing the residual variability by one or two percent is unlikely to have any real benefit.

Variables that are not included in these three categories are those that are unrelated to the response variable and so are irrelevant, and those that are affected by the exposure of interest or by the response, and so would mask the relationship being studied.

Ideally, all the variable selection needed for a particular analysis can done before getting the data. This does not imply that only a single model should be fitted, as it can be interesting to find out which confounders have a big impact and which have a small impact. Since the choice of confounders and primary exposure variables determines what the coefficients mean, it is undesirable for this choice to be random, depending on the values in one particular sample. Model selection based on the data is sometimes a necessary evil, when there are too many variables that might be confounders and insufficient data to include them all in the same model.

This view of model selection in investigating associations is in contrast to the situation where models for prediction are being constructed. A flexible model selection procedure together with a reliable estimate of predictive accuracy can give accurate models for predicting Y at new values of X, but will likely not result in the right confounders being included to give the coefficients a useful interpretation. Predictive models, such as the Framingham risk score for cardiovascular disease or actuarial risk scores in the insurance and credit industries, are not usually constructed from complex survey data, and will not be discussed further here.

5.2.4   Linear models in the survey package

The function svyglm() fits linear and generalized linear models to data stored in a survey design object. The function syntax is almost identical to the model-based glm(), except that the data argument of glm() is replaced by a design argument.

The main difference between svyglm() and glm() output is that the sampling-weighted least squares used by svyglm() is not maximum likelihood. Even if the model were assumed to be true it would not be possible to compare models using likelihood ratio tests, and the models fitted by svyglm() do not have methods for the generic likelihood-ratio test function anova(). In principle it is possible to extend the Rao–Scott tests from 2×2 tables to regression models, but this has been implemented only for the log-linear models in Section 6.3.

The summary() function, applied to a svyglm() object, will give Wald tests for each coefficient in the model. For a collection of coefficients, such as the coefficients for a multi-category factor variable, use regTermTest() to compute Wald tests.

Example: Dietary sodium and potassium and blood pressure.   High salt diets (and other sources of sodium) are known to increase blood pressure, although the details of the relationship are complicated and some aspects are controversial. We will examine the association between sodium and potassium consumption and blood pressure in data from NHANES 2003–2006. This data set consists of two two-year waves of the new "continuous" NHANES design, so it is necessary to download data on demographics, diet, body size, and blood pressure, for both NHANES 2003–2004 and NHANES 2005–2006, extract the appropriate variables, and merge the data sets.

Figure 5.9 shows code for loading the eight files and merging them. Within each wave, merge() is used to combine the four data files, and then rbind() is used to paste the two results data sets together vertically (of course, after checking that the variables do in fact line up correctly). It is also necessary to compute the variables BPXSAR and BPXDAR, averages of the multiple blood pressure measurements, which are provided in the 2003–2004 data but not in 2005–2006.

The weights need to be adjusted for the combined data. Since each wave of analysis is weighted to correspond to the full US population, the combined data would represent two copies of the population. A new fouryeartwt variable is created by halving the weight variable WTDRD1 that is recommended for analysis of the dietary data. In fact, since we will be estimating only means, not totals, it would have made no difference to the regression results if WTDRD1 had been used as the weights, without rescaling. Finally, the survey design object is created, using the recommended single-stage approximation to the NHANES design.

Figure 5.10 shows the systolic blood pressure and dietary sodium data, with a smooth mean curve produced by svysmooth(). There is no clear relationship, and in fact a suggestion that blood pressure decreases as sodium intake increases. Recalling the graphs of blood pressure and age from Chapter 4, it is likely that age

```
demo <- read.xport("demo_c.xpt")[, c(1:8,28:31)]
bp <- read.xport("bpx_c.xpt")
bm <- read.xport("bmx_c.xpt")[, c("SEQN","BMXBMI")]
diet <- read.xport("dr1tot_c.xpt")[, c(1:52,63,64)]

nhanes34 <- merge(demo, bp, by="SEQN")
nhanes34 < -merge(nhanes34, bm, by="SEQN")
nhanes34 <- merge(nhanes34, diet, by="SEQN")

demo5 <- read.xport("demo_d.xpt")[,c(1:8,39:42)]
bp5 <- read.xport("bpx_x.xpt")
bp5$BPXSAR <- rowMeans(bp5[,c("BPXSY1","BPXSY2","BPXSY3",
    "BPXSY4")], na.rm=TRUE)
bp5$BPXDAR <- rowMeans(bp5[,c("BPXDI1","BPXDI2","BPXDI3",
    "BPXDI4")], na.rm=TRUE)
bm5 <- read.xport("bmx_d.xpt")[,c("SEQN","BMXBMI")]
diet5 <- read.xport("dr1tot_d.xpt")[,c(1:52,64,65)]

nhanes56 <- merge(demo5,bp5,by="SEQN")
nhanes56 <- merge(nhanes56,bm5,by="SEQN")
nhanes56 <- merge(nhanes56,diet5,by="SEQN")

nhanes <- rbind(nhanes34,nhanes56)
nhanes$fouryearwt <- nhanes$WTDRD1/2

des <- svydesign(id=~SDMVPSU,strat=~SDMVSTRA,weights=~fouryearwt,
    nest=TRUE, data=subset(nhanes, !is.na(WTDRD1)))
des <- update(des, sodium=DR1TSODI/1000, potassium=DR1TPOTA/1000)
```

Figure 5.9 Loading and merging data from NHANES 2003–2004 and NHANES 2005—2006

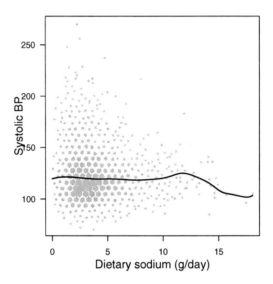

Figure 5.10   Dietary sodium and systolic blood pressure, from NHANES 2003–2006

is an important confounder. Blood pressure certainly increases with age, and it is quite likely that people of different ages have different diets. Figure 5.11 shows the relationship between blood pressure and dietary sodium for six age groups, and there is now a suggestion of an increase, especially at younger ages.

A regression model with just sodium and potassium as predictors can be fitted by

```
model0 <- svyglm(BPXSAR~sodium+potassium, design=des)
```

The summary() method for model0 gives a table with standard errors and p-values. This gives point estimates of −0.69 mmHg/g/day for sodium and 0.78 for potassium, with respective p-values of 0.0002 and 0.0068. These estimates are in the wrong direction, like the graph in Figure 5.10, because of confounding by age. Adjusting for age

```
model1 <- svyglm(BPXSAR~sodium+potassium+RIDAGEYR, design=des)
```

reverses the association, with coefficients of 0.80 for sodium and −0.91 for potassium and p-values less than 0.0001. Other important potential confounding variables are gender (RIAGENDR) and some measure of obesity such as BMI (BMXBMI). Table 5.1 shows the coefficients for sodium and potassium for both systolic and diastolic blood pressure in a range of models.

A combined test for the effects of sodium and potassium can be computed with regTermTest() as follows

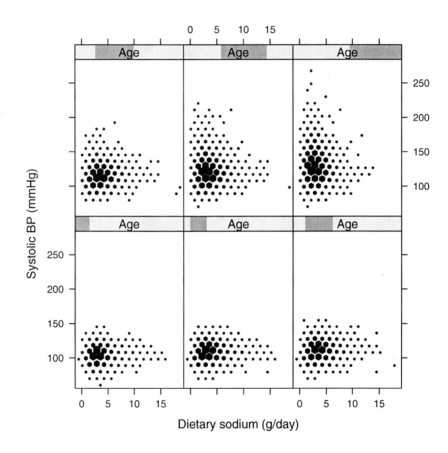

Figure 5.11   Dietary sodium and systolic blood pressure, by age, from NHANES 2003–2006

Table 5.1 Associations between dietary sodium and potassium and blood pressure in NHANES 2003–2006. The units for all coefficients are mmHg/g/day. Standard errors are given in parentheses

Model	Systolic BP		Diastolic BP	
	Sodium	Potassium	Sodium	Potassium
Unadjusted	−0.69	0.78	−0.05	0.88
	(0.10)	(0.19)	(0.10)	(0.21)
Age and gender	0.59	−1.09	0.34	0.26
	(0.16)	(0.18)	(0.10)	(0.19)
Age, gender, BMI	0.43	−0.96	0.34	0.25
	(0.16)	(0.17)	(0.10)	(0.19)

```
> regTermTest(model2, ~potassium+sodium, df=NULL)
Wald test for potassium sodium
 in svyglm(BPXSAR~sodium+potassium+RIDAGEYR+RIAGENDR+BMXBMI,
    design = des)
F =  15.98481  on  2  and  25  df: p= 3.3784e-05
```

where df=NULL gives the default choice for residual degrees of freedom, which in this case is the number of strata minus the number of PSUs in the design.

Although the associations are statistically significant they are not very strong: a gram of sodium is about 2.5 grams of salt, which is quite a lot, and a 0.43 mmHg difference in blood pressure is very small. It is possible that a larger effect is being missed for various reasons: there is substantial measurement error in dietary data, the effect might be nonlinear, it might depend on history of salt consumption rather than just current consumption, or it might depend on age. Some of these possibilities can be examined with the available data.

The left-hand panel of Figure 5.12 shows the residuals and fitted values from the final model for systolic blood pressure, with a smooth mean curve estimated with svysmooth(). Because some variables have missing values it is necessary to subset the survey design object to include just the non-missing subpopulation for the graph and smooth curve. The model object has an element na.action that lists the observation numbers that were dropped because of missing data, so des[-model1$na.action,] is the subsample of design without these missing values. There is no suggestion of nonlinearity in the fitted values that might indicate misspecification of the confounding model. The right-hand panel shows partial residuals for sodium. Partial residuals are obtained by adding back the fitted contribution of one variable to the residuals, in this case

$$\text{partial residual} = \text{residual} + 0.43 \times \texttt{sodium},$$

and can be computed with resid(model1, "partial"). If the model is misspecified, the mean of the partial residuals will not follow the fitted relationship, and this can be detected using a smooth mean curve. In the graph, the solid line is the fitted

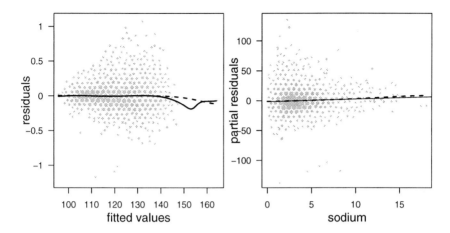

Figure 5.12   Diagnostic plots. The left-hand panel shows scaled residuals vs fitted values, and smooth mean curves with bandwidths of 5 and 10 mmHg. The right-hand panel shows partial residuals for sodium intake, with the fitted straight line (solid) and a smooth curve(dashed).

relationship, a straight line with slope 0.43. The dashed curve is a fitted mean using svysmooth(). There is no suggestion of nonlinearity in the relationship between sodium and blood pressure.

With unweighted data these and other diagnostic plots can be produced with the plot() method and by the termplot() function. These functions will also work for models fitted by svyglm(), but by default the smooth curves they fit will not use the sampling weights. To use termplot() and plot() with sampling-weighted models it is possible to supply a weighted smoother that slots in to replace the usual unweighted smoother

```
nonmissing<-des[-model1$na.action]
plot(model1, panel=make.panel.svysmooth(nonmissing))
termplot(model1, data=model.frame(nonmissing),
  partial=TRUE, se=TRUE,
  smooth=make.panel.svysmooth(nonmissing))
```

The plots will still not show the sampling weights for the plotted points, but these functions provide a simple way of obtaining initial versions of the diagnostic plots.

It is possible to construct diagnostic plots to assess interactions, but it is much easier just to fit the models. The model

```
int1<-svyglm(BPXSAR ~ (sodium + potassium)*I(RIDAGEYR-40)
  +RIAGENDR + BMXBMI, design = des)
```

adds interaction terms between age and each of sodium intake and potassium intake. The reason for using RIDAGEYR-40 is so that the zero of the age variable occurs at a useful age rather than at zero age. The asterisk in the model requests both the main effect of age and the interaction term. The coefficients in this model and the

Table 5.2 Interactions between age and sodium and potassium intake, data from NHANES 2003–2006

	Systolic BP			Diastolic BP		
	Coefficient	Std Error	*p*-value	Coefficient	Std Error	*p*-value
Sodium	0.310	0.166	0.074	0.310	0.114	0.012
Potassium	-0.976	0.182	< 0.001	0.326	0.182	0.087
Sodium:Age	−0.016	0.009	0.086	0.045	0.007	< 0.001
Potassium:Age	−0.040	0.010	< 0.001	−0.037	0.007	< 0.001

analogous one for diastolic pressure are given in Table 5.2 With this coding for age, the main effects in Table 5.2 are the associations at age 40, and the interactions show how this changes for each additional year of age. The association between potassium and lower blood pressure strengthens with age, as does the association between sodium and higher diastolic pressure. The association between sodium and higher systolic pressure appears stronger in the young. Even with these trends, the estimated associations are not very strong. At age 80, the interaction model gives a coefficient of −2.6 mmHg/g/day for potassium on systolic pressure and −1.2 mmHg/g/day for diastolic pressure.

Measurement error is a important possibility, since accurate assessment of diet is notoriously difficult. A second dietary interview was conducted later, by telephone, and looking at the 2005–2006 data for the second interview shows that the differences in reported sodium and potassium intakes were often large. The variance of the difference was 1.3 g/day for potassium and 3.6 g/day for potassium (computed with svyvar()), and on average the second interview reported lower sodium consumption (by 0.14 g/day). These results suggest that more than half the measured person-to-person variability in sodium intake is attributable to measurement error, and this is before considering any bias in the measurement process. It is quite possible that this measurement error is partially masking a large effect of dietary sodium and potassium on blood pressure.

5.3 IS WEIGHTING NEEDED IN REGRESSION MODELS?

Sampling weights are important in design-based analyses because unequal sampling can distort associations between variables, and this must be corrected in the analysis. Since regression models use adjustment for confounders as a way of removing distorted associations between exposure and response, it is plausible that a regression model might not need sampling weights. Another way of making the same argument is that ignoring the sampling weights gives a regression model fitted to a population with a different distribution of some variables from that in the real population. Since model-based regression is primarily used to uncover associations that behave stably across populations, it should be possible to estimate these associations without using sampling weights. An examination of the possible biases for a range of assumptions about true relationships is given by DuMouchel and Duncan [43].

Fitting regression models without sampling weights has been attractive for two reasons. The first is that not all software can use sampling weights; this is much less important today. The second reason, still relevant now, is that weighting will typically reduce precision. If the sampling weights are *ignorable*, in the sense that the estimate is valid with or without the weights, the weighted estimate will be less precise.

There are two main limitations to the ability of regression models to adjust for biased sampling without using the weights. The first is that some important variables used in constructing the weights may not be available, the second is that they may not be suitable variables to include in the model. Consider the California Health Interview Survey. The weights depend on age, sex, and race/ethnicity, and also on the total number of people, number of children, and number of telephones in a household. The CHIS data are suitable for examining associations between socioeconomic variables and health, and in this context a regression model would likely want to adjust for age, sex, and race/ethnicity. It might well be undesirable to adjust for household size, family size, and number of telephones, since these could well be affected by the socioeconomic variables being examined. Even in situations where there is no obvious reason for the unweighted regression estimate to be biased it is still wise to be cautious. A relatively small bias is sufficient to outweigh the reduction in standard error and give an overall less-accurate estimate, and biases of this size cannot reliably be detected from the data.

Fortunately, as long as model-robust standard error estimates are used to account for clustering and for non-constant variance it often makes little practical difference whether sampling weights are used in fitting regression models. For example, the final model for systolic blood pressure, sodium, and potassium in Table 5.1 gives coefficients of 0.43 and -0.96 for sodium and potassium, with standard errors of 0.16 and 0.17; an unweighted linear model gives 0.35 and -0.99 as the coefficients, with standard errors of 0.09 and 0.13. There is no real difference in the conclusions that would be drawn from these analyses. When there is a substantive difference it may indicate that a few influential observations happen to have large sampling weights, so that neither the weighted nor the unweighted analysis is entirely reliable. Refitting the models after removing the most influential observations is a useful sensitivity analysis.

The use of sampling weights in case–control designs is a separate issue, and is discussed in section 8.3.

EXERCISES

5.1   This exercise uses the Washington State Crime data for 2004 as the population. The data consist of crime rates and population size for the police districts (in cities/towns) and sheriffs' offices (in unincorporated areas), grouped by county.

a) Take a simple random sample of ten counties from the state and use all the data from the sampled counties. Estimate the total number of murders and of burglaries in the state.

b) Use the population of each county as an auxiliary variable to estimate the totals.

c) Use the numbers of murders and burglaries in the previous year as auxiliary variables in a regression estimate of the totals (why can't we use a ratio estimate here?).

d) Stratify the sampling so that King County is sampled with 100% probability together with a simple random sample of five other counties. Use population as an auxiliary variable to construct a common ratio estimate and a separate ratio estimate (section 5.2.2) of the population totals.

e) Take simple random samples of five police districts from King County and five counties from the rest of the state. Use population as an auxiliary variable to construct a common ratio estimate and a separate ratio estimate (section 5.2.2) of the population totals

5.2   Using the Washington State Crime data as a population, take a stratified sample of five police districts from King County and five counties from the rest of the state. Estimate the ratio of violent crimes to non-violent crimes. Compare to the population value.

5.3   Using the data from Wave 1 of the 1996 SIPP panel (see Figure 3.8),

a) Estimate the ratio of population totals for monthly rent (`tmthrnt`) and total household income (`thtrninc`) over the whole population and over the subpopulation who pay rent.

b) Compute the individual-level ratio, i.e., the proportion of household income paid in rent, and estimate the population mean over the whole population and the subpopulation who pay rent.

5.4   Use the stratified sample from the Academic Performance Index population to examine whether the proportion of teachers with only emergency qualifications (`emer`) affects academic performance (measured by 2000 API).

a) What confounding variables measuring socioeconomic status of students should be included in the model?

b) Should 1999 API be in the model (and why, or why not)?

c) Do any of the confounding variables need to be transformed?

d) Does `emer` need to be transformed?

e) What is the conclusion at the end?

5.5   Following on from the previous exercise, fit the same model to the whole population (the data set `apipop`) using the `glm()` function.

a) Do the sample estimates agree with the population data? Do your decisions about transforming variables hold up in the population data?

b) ⋆ Fit the same model to 100 stratified samples from the population (hint: use `replicate()`). Is the sampling distribution of the coefficients close to a Normal distribution?

5.6   Using the blood pressure data from NHANES 2003–2006, investigate the effect of obesity on blood pressure using the Body Mass Index and blood pressure data.

   a) What variables in the data set are potential confounders?

   b) Are there important confounders that are not measured?

   c) Fit one or more suitable regression models and summarize the output.

   d) Examine whether there is an interaction with age or sex.

5.7   ⋆ Prove that an unweighted regression estimator is approximately unbiased when the weights depend only on variables in the model. Specifically, if the true population regression coefficients β^* satisfy

$$\sum_{i=1}^{N} x_i (y_i - x_i \beta^*) = 0$$

and R_i indicates that observation i is in the sample prove that

$$E\left[\sum_{i=1}^{N} R_i x_i (y_i - x_i \beta^*)\right] = 0$$

so that the unweighted sample estimating equations are unbiased.

5.8   ⋆ A rough approximation to the loss of efficiency from unnecessarily using weights can be constructed by considering the variance of the residuals in weighted and unweighted estimation. Assume as an approximation that the residuals r_i are independent of the sampling weights w_i

   a) Show that

$$\text{var}\left[\sum_{i=1}^{n} w_i r_i\right] = E[w^2]\text{var}[r] + \text{var}[w]E[r^2].$$

   b) Now assume that the mean of the residuals is zero, and show that the relative efficiency of the unweighted estimate is $1 + \text{cv}(w)$, where cv is the coefficient of variation, the ratio of the standard deviation to the mean.

CHAPTER 6

CATEGORICAL DATA REGRESSION

In which trends become apparent.

This chapter discusses analytic techniques for binary and categorical data. The first two sections describe regression models where the mean of the outcome variable is a proportion or set of proportions. If there is a strong relationship between between a continuous predictor and a proportion, the relationship cannot be linear, since the proportion must be between zero and one, so a transformation is needed to link the predictor and the proportion. Survival models for risk of an event over time, which are structurally very similar, are deferred until Chapter 8, where we will have real data to apply them to. The principles for constructing and interpreting these regression models for proportions are the same as for the linear regression models in the previous chapter, with the exception that useful precision variables are almost non-existent outside linear regression. The final section of this chapter covers loglinear models, which are used to describe relationships among a set of categorical variables, without necessarily specifying one as an outcome and the others as predictors.

Other generalized linear models (e.g., Poisson regression for count data or Gamma regression for positive, highly skewed, continuous variables) can be fitted with

Complex Surveys: A Guide to Analysis Using R. By Thomas Lumley
Copyright © 2010 John Wiley & Sons, Inc.

svyglm(), but are not discussed in detail here. It may also sometimes be necessary to fit regression models that are not built in to the survey package. Appendix E gives a case study of implementing a model (negative binomial regression) where the outcome variable is a count of occurrences.

6.1 LOGISTIC REGRESSION

Logistic regression is probably second only to linear regression in frequency of use. It has been the foundation of statistical analysis in epidemiology, partly for reasons that are discussed in Section 8.3, but is also widely used in others areas of the health sciences and in econometrics.

The logistic regression model for a binary response variable Y and predictor variables $X_1, X_2, \ldots, X_p$ is

$$\mathrm{logit}\, P[Y = 1] = \alpha + \beta_1 X_1 + \beta_2 X_2 + \cdots + \beta_p X_p$$

where the logit function is defined as

$$\mathrm{logit}(p) = \log\left(\frac{p}{1-p}\right).$$

The logit of $P[Y = 1]$ is also the logarithm of the *odds* of $Y = 1$, where the odds is defined as $P[Y = 1]/P[Y = 0]$.

The parameters β are now average differences in $\mathrm{logit}\, P[Y = 1]$ for a one-unit difference in X, rather than average differences in the mean of Y as in linear regression. If β is a difference in the logarithm of the odds of $Y = 1$ for one-unit differences in X, then e^β is a ratio of the odds of $Y = 1$ for a one-unit difference in X. The results of logistic regression are most often described in terms of e^β, the *odds ratio*.

Logistic regression models are fitted using svyglm(). The model formula and the survey design object are specified in the same way as for linear regression. To specify logistic regression, rather than the default of linear regression, use the option family=quasibinomial().

Example: Internet use in Scotland.   A subset of variables from the Scottish Household Survey have been made available (after some editing for confidentiality) by the PEAS project at Napier University, and the analyses here are similar to those on the PEAS web site. The data set contains information on internet use, income, age, sex, and region of Scotland. The survey is a slightly unusual design, in that it use a stratified sample of individuals in areas with high population density and a stratified cluster sample in areas of low population density. The aim of this design is to use cluster-sampling, which is less efficient, only in areas where it is unavoidable. No special handling is needed to specify this design to svydesign(). The PSUs are given in the data set, and it does not matter that some of them are individuals and some are clusters.

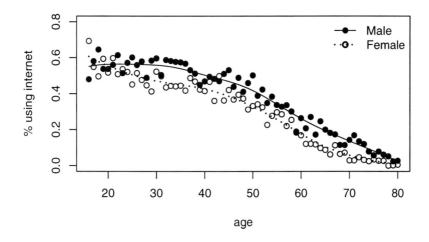

Figure 6.1   Proportion using the internet in Scotland by age and sex, based on Scottish Household Survey 2001–2

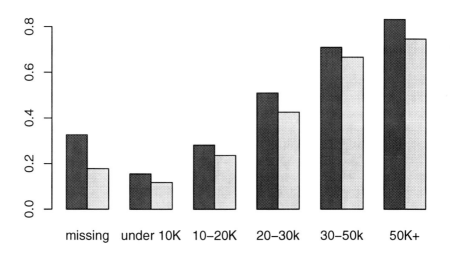

Figure 6.2   Proportion using the internet in Scotland by income and sex, 2001–2, based on Scottish Household Survey 2001-2. Shaded bars are proportions for men.

```
shs<-svydesign(id=~psu, strata=~stratum, weight=~grosswt,
    data=shs_data)
bys<-svyby(~intuse,~age+sex,svymean,design=shs)
plot(svysmooth(intuse~age,
        design=subset(shs,sex=="male" & !is.na(age)),
    bandwidth=5),ylim=c(0,0.8),ylab="% using internet")
lines(svysmooth(intuse~age,
        design=subset(shs,sex=="female" & !is.na(age)),
    bandwidth=5),lwd=2,lty=3)
points(bys$age,bys$intuse,pch=ifelse(bys$sex=="male",19,1))
legend("topright",pch=c(19,1),lty=c(1,3),lwd=c(1,2),
    legend=c("Male","Female"),bty="n")
byinc<-svyby(~intuse, ~sex+groupinc, design=shs)
barplot(byinc)
```

Figure 6.3 Computing the proportion using the internet in Scotland

Scatterplots are not very useful for binary data, so Figure 6.1 shows smooth estimated mean curves (computed with svysmooth()) and estimated means for each sex:year-of-age combination, and Figure 6.2 shows a bar plot of internet use by income and sex. The code for these graphs is in Figure 6.3. Women are less likely to use the internet than men, except at the youngest ages, and internet use falls off with age and increases with income.

The lower level of internet use at higher age illustrates why it is important to be careful in describing and interpreting regression coefficients. It is almost certainly not the case that as people get older they will be less likely to use the internet. If the same people were surveyed now, seven years later, we would expect that almost everyone who had been an internet user still was, and that many non-users had started. The relationship with age is a *cohort effect*, not an age effect: people who were 60 years old in 2001 were born in 1941 and grew up before public internet access existed. The trend with increasing income, on the other hand, probably does reflect a real effect. Someone who previously had a low income and begins a job earning £30,000 is quite likely to start using the internet.

Figure 6.1 also shows a difference in the shape of the age relationship for men and women. For men, the curve is roughly flat through the mid-thirties, for women it begins to descend immediately. This fits with the cohort-effect interpretation if, historically, men started using the internet is large numbers before women did, but it could also be an effect of income or of different types of jobs.

Figure 6.4 shows code and output for fitting logistic regression models to the data. The first model, m, has an age by sex interaction, that is, it fits a separate slope in age for men and for women. Age is represented by the variable I(age-18) because the coefficient for sex will be the difference between men and women when the age variable is zero. Subtracting 18 from age makes this comparison between men and women occur at a meaningful age within the range of the data. The output produced by summary(m) shows the coefficients β, not the odds ratios. Negative coefficients correspond to odds ratios less than 1.0, and positive coefficients to odds ratios greater than 1.0.

There is very strong evidence of a strong association with age. The odds ratio is $\exp(-0.4497) = 0.96$, so the odds of a man being an internet user is about 4% lower for each year of age. There is also strong evidence of a difference in slope between men and women, with the decrease in odds of internet use being faster for women than for men. There is weak evidence of a difference between men and women at age 18.

According to Figure 6.1, the difference in the slope with age between men and women is largely at earlier ages, where there the slope is close to zero in men. Model m2 fits a *linear regression spline* to age. That is, the age relationship below 35 and above 35 are lines with different slopes meeting at age 35. The functions pmin() and pmax() compute element-wise minimum or maximum of two vectors, so pmin(age, 35) is the age, for ages up to 35, and is 35 for ages above 35. A one-unit difference in this variable is a comparison between two ages that differ by one year and are both below 35. Similarly, a one-unit difference in pmax(age, 35) compares two ages that differ by one year and are both above 35. The coefficients are the average slopes for ages up to 35 and over 35. Although a broken-line relationship of this sort is not a plausible candidate for a precisely true model, it is very useful in producing interpretable summaries of relationships.

The coefficient of pmin(age, 35), giving the slope for men at ages up to 35, is positive but small (although statistically significant). The slope after age 35 is negative and large, matching the impression in Figure 6.1 of a roughly flat relationship at earlier ages and then a steady decline. Working out the two slopes for women requires adding the slope for men and the interaction term, which is done using svycontrast(). The slope for women is negative but small up to age 35, and negative and large above age 35. That is, both sexes show a flatter relationship with age below 35 and a steeper relationship above age 35.

Income is the next predictor to explore. The variable groupinc is inconveniently coded so that the first level, which is used as the reference level in the default coding, is "missing". The code in Figure 6.5 starts by defining a new income variable whose reference level is the lowest income group. Fitting the model shows there is still a strong age trend, and now a strong income trend. The differences between men and women are now smaller: women on average earn less, so that of the difference between men and women can be explained by differences in income.

Model m4 includes an interaction between sex and income, i.e., a different income trend for men and women. If we do not have any particular interest in estimating these trends separately it is worth testing whether there is evidence of a difference. The interaction term adds five extra parameters, so looking at test of single parameters will not be helpful. With regTermTest() we test the hypothesis that all five of the added variables are zero, and there is little evidence against this hypothesis. The Wald test has been found to be unreliable in complex designs with a small number of PSUs, as discussed for contingency tables in section 2.4.2, but the Scottish Household Survey has a very large number of small PSUs and there should not be a problem. It would

```
> m<-svyglm(intuse~I(age-18)*sex,design=shs,
    family=quasibinomial())
> m2<-svyglm(intuse~(pmin(age,35)+pmax(age,35))*sex,
    design=shs,family=quasibinomial)
> summary(m)
Call:
svyglm(intuse ~ I(age - 18) * sex, design = shs,
    family = quasibinomial())

Survey design:
svydesign(id = ~psu, strata = ~stratum, weight = ~grosswt,
    data = ex2)

Coefficients:
                         Estimate Std. Error t value Pr(>|t|)
(Intercept)              0.804113   0.047571  16.903  < 2e-16 ***
I(age - 18)             -0.044970   0.001382 -32.551  < 2e-16 ***
sexfemale               -0.116442   0.061748  -1.886   0.0594 .
I(age - 18):sexfemale   -0.010145   0.001864  -5.444 5.33e-08 ***
> summary(m2)
Call:
svyglm(intuse ~ (pmin(age, 35) + pmax(age, 35)) * sex,
    design = shs, family = quasibinomial)

Survey design:
svydesign(id = ~psu, strata = ~stratum, weight = ~grosswt,
    data = ex2)

Coefficients:
                            Estimate Std. Error t value Pr(>|t|)
(Intercept)                 2.152291   0.156772  13.729  < 2e-16 ***
pmin(age, 35)               0.014055   0.005456   2.576 0.010003 *
pmax(age, 35)              -0.063366   0.001925 -32.922  < 2e-16 ***
sexfemale                   0.606718   0.211516   2.868 0.004133 **
pmin(age, 35):sexfemale    -0.017155   0.007294  -2.352 0.018691 *
pmax(age, 35):sexfemale    -0.009804   0.002587  -3.790 0.000151 ***
> svycontrast(m2,
    quote('pmin(age, 35)' +'pmin(age, 35):sexfemale'))
            nlcon     SE
contrast -0.0031 0.0049
> svycontrast(m2,
    quote('pmax(age, 35)' + 'pmax(age, 35):sexfemale'))
            nlcon     SE
contrast -0.07317 0.0018
```

Figure 6.4 Fitting logistic regression models for internet use and age, to data from the Scottish Household Survey

```
> shs<-update(shs,income=relevel(groupinc, ref="under 10K"))
> m3<-svyglm(intuse ~ (pmin(age, 35) + pmax(age, 35)) * sex+income,
    design = shs,family = quasibinomial)
> summary(m3)
Call:
svyglm(intuse ~ (pmin(age, 35) + pmax(age, 35)) * sex + income,
    design = shs, family = quasibinomial)
Survey design:
update(shs, income = relevel(groupinc, ref = "under 10K"))
Coefficients:
                          Estimate Std. Error t value Pr(>|t|)
(Intercept)               1.275691   0.179902   7.091 1.41e-12 ***
pmin(age, 35)            -0.009041   0.006170  -1.465  0.14286
pmax(age, 35)            -0.049408   0.002124 -23.259  < 2e-16 ***
sexfemale                 0.758883   0.235975   3.216  0.00130 **
incomemissing             0.610892   0.117721   5.189 2.15e-07 ***
income10-20K              0.533093   0.048473  10.998  < 2e-16 ***
income20-30k              1.246396   0.052711  23.646  < 2e-16 ***
income30-50k              2.197628   0.063644  34.530  < 2e-16 ***
income50K+                2.797022   0.132077  21.177  < 2e-16 ***
pmin(age, 35):sexfemale  -0.023225   0.008137  -2.854  0.00432 **
pmax(age, 35):sexfemale  -0.008103   0.002858  -2.835  0.00459 **
> m4<-svyglm(intuse~(pmin(age, 35)+pmax(age, 35))*sex+income*sex,
    design = shs,family = quasibinomial)
> regTermTest(m4, ~income:sex)
Wald test for income:sex
 in svyglm(intuse~(pmin(age, 35)+pmax(age, 35)) * sex + income *
    sex, design = shs, family = quasibinomial)
Chisq = 9.364053  on  5  df: p= 0.095395
```

Figure 6.5   Adding income to predictors of internet use in Scotland

be desirable to have tests similar to the Rao–Scott tests of sections 2.4.2 and 6.3, and these are likely to be implemented in the future. In this example the Rao–Scott test gives a p-value of 0.11, not meaningfully different from the 0.095 of the Wald test. Code for computing the Rao–Scott test in this example is on the web site.

A final observation on these data is that they illustrate the difference between linear and logistic models in the benefit from precision variables. Age and sex are approximately independent, with a weak association due to longer life expectancy for women. Since age is very strongly associated with internet use, adjusting for age might be expected to increase the precision for estimating the association between sex and internet use. In a linear regression there would be an increase in precision, but in a logistic regression things are more complicated. The coefficient for sex in a model with no other variables is is -0.346 with standard error 0.029. In a model with formula `intuse~sex+age` the coefficient is -0.377 with standard error 0.32. The standard error has increased slightly, as has the coefficient, and the z-statistic for testing the hypothesis of no association with sex is essentially unchanged. The change in the coefficient is in the opposite direction to what would be expected from the association between age and sex: older people are more likely to be women, so in the unadjusted model the sex variable would capture a small amount of the age association and show a larger effect.

The increase in coefficient and standard error, with little impact on the z-statistic is typical of logistic regression (and Cox regression). Because of the peculiarities of odds ratios, the adjusted and unadjusted odds ratio are not the same even when predictor variables are exactly independent of each other. This phenomenon, called *non-collapsibility*, is discussed in more detail by Greenland et al. [53].

6.1.1   Relative risk regression

When $P[Y = 1]$ is small, $\text{logit} P[Y = 1]$ is very close to $\log P[Y = 1]$, and the odds ratios estimated by logistic regression are close to risk ratios. If we find an odds ratio of 10 for lung cancer comparing smokers and non-smokers, it is true to say that smokers were 10 times as likely as non-smokers to get lung cancer. If $P[Y = 1]$ is not small, the odds ratio will be larger (further from 1.0) than the relative risk, and confusing the two summaries can be seriously misleading.

The Scottish Household Survey analyses in the previous section illustrate this problem. The odds ratio comparing the highest and lowest income groups is $\exp(2.797) = 16.4$. It is tempting to interpret this as 16.4 times higher proportion using the internet. Since the proportion of internet users in the lowest income group is 15% for men and 12% for women, the proportion in the highest income group would then be 192% for women and 246% for men, which is clearly wrong. It is quite difficult to stop this interpretation being used, as was illustrated by a paper in the New England Journal of Medicine reporting a well-designed experimental study of race and gender bias in referral for cardiac catheterization[155, 157]. Although the results were reported correctly in the article, the press release and subsequent media coverage described an odds ratio of 0.6 as 40% lower chance of referral, when the actual referral ratio was 0.93.

It is possible to fit a regression model where the regression parameters estimate population log relative risks instead of population log odds ratios, by specifying a different *link function* in `family` argument to `svyglm()`. Logistic regression uses the logit link function, i.e., it fits a linear relationship to $\text{logit} P[Y = 1]$. Relative risk regression replaces this with the logarithm, fitting a linear relationship to $\log P[Y = 1]$.

One approach in R is to use `family = quasibinomial(log)`, another is to use `quasipoisson(log)`. Both approaches estimated a population average relative risk, but they give different weights to outlying predictor values. Using `quasibinomial(log)` requires the fitted probabilities for the working model to be less than 1.0, and `quasipoisson(log)` does not. Since the model is just a working model the arguments against having fitted probabilities greater than 1.0 are weaker than for model-based inference, but this is an issue that attracts much controversy [2, 39, 102]. When using `quasipoisson(log)` it would be prudent to check that the number of fitted values greater than 1.0 is very small, and that removing these observations does not have a large impact on the coefficients.

Using `quasibinomial(log)` is computationally more challenging. It requires specifying starting values to `svyglm()` and it may also be necessary to increase the number of iterations by specifying, eg, `maxit=100`. A fairly reliable set of starting values is -0.5 for the intercept and zero for other coefficients. Fitting relative risk versions of model m3 from the previous section:

```
rr3<-svyglm(intuse~(pmin(age, 35) + pmax(age, 35))*sex+income,
    design = shs, family = quasibinomial(log),
    start=c(-0.5,rep(0,10)))
rr4<-svyglm(intuse~(pmin(age, 35) + pmax(age, 35))*sex+income,
    design = shs, family = quasipoisson(log))
```

gives a relative risk for the highest vs lowest income categories of 3.5 in model `rr3` and 3.9 in model `rr4`. These are fairly similar and are very different from the odds ratio of 16.4. The model using `family = quasipoisson(log)` has fitted values above 1.0 for 0.5% of observations, young people with very high incomes. These people are virtually all internet users and their fitted values are not much greater than 1.0. In this case either model is reasonable.

6.2 ORDINAL REGRESSION

A variable with an ordered set of K categories could be turned into a binary variable by dichotomizing at any of the $K - 1$ breaks between categories. Dichotomizing at $Y \leq k$ vs $Y > k$ would give a logistic regression model

$$\text{logit} \Pr[Y > k] = \alpha_k + x\beta_k. \tag{6.1}$$

In the $K - 1$ models arising from the $K - 1$ possible choices of k it is possible that β_k is approximately the same, but it is impossible for α_k to be the same since α_k is the log odds of $Y > k$ at $x = 0$. A simplified model could then be

$$\text{logit } \Pr[Y > k] = \alpha_k + x\beta. \tag{6.2}$$

Equation 6.2 defines the *proportional odds model* or *ordinal logistic model* of Mc-Cullagh [105]. The parameter estimates are log odds ratios for one-unit differences in x averaged over observations and over cutpoints k.

Example: Urinary cadmium and diabetes Schwartz et al. [156] analyzed data from NHANES III and found that cadmium concentration in the urine was associated with a higher risk of diabetes and elevated fasting plasma glucose. They performed an unweighted analysis; we will consider a weighted analysis of a slightly different subset of the data.

Fasting glucose (as the name indicates) is measurable only if the participant has not eaten for eight hours before blood draw. This is easier to achieve in the morning and approximately 90% of the morning laboratory data are in the fasting state. The NHANES data come with a set of weights that are specific to the morning interview session, so we will use only the morning data. The upper panel of Figure 6.6 shows the data on urinary cadmium and fasting plasma glucose as a hexagonally binned scatterplot with a set of smooth curves for the 10th, 25th, 50th, 75th, 90th, and 95th quantiles. It does appear that the higher quantiles of fasting glucose are increased in people with higher urinary cadmium (the decrease at the extreme right is due to a shortage of data). The lower panel follows Schwartz et al. in using the ratio of cadmium to creatinine rather than the cadmium measurement. This is intended to correct for differences in urine concentration due to differences in hydration or in renal function. It also runs the risk of introducing spurious signals from unusually low or high creatinine levels. The association seems stronger in this second graph. Code for these graphs is on the web site — adding curves to a hexagonal-bin plot is relatively complex.

Fasting glucose is typically divided into three categories, though the standard cutpoints for these categories have evolved over time. With the American Diabetic Association categories that were in use at the time of Schwartz's analysis, fasting plasma glucose concentrations below 110 mg/dl are considered normal, those above 125mg/dl indicate diabetes, and everyone with previously diagnosed diabetes is placed in this category regardless of their glucose level. The range 110–125 mg/dl indicates impaired metabolism of glucose, with an increased risk of future diabetes. The current version of the criteria uses 100 mg/dl rather than 110 mg/dl as the threshold for impaired metabolism and the results of the analysis are similar using the modern categories.

It would be possible to model the three-level category using two logistic regression models, for example, one with diabetes as the binary outcome and one with glucose above or below 110 mg/dl as the outcome. The proportional odds model combines these two into a single summary model. Figure 6.7 shows the definition of variables and the models. Rather than use the ratio of cadmium to creatinine these models include both variables. If the association is entirely through the ratio, the coefficient for creatinine will be −0.01 times the coefficient for cadmium (the multiplier would

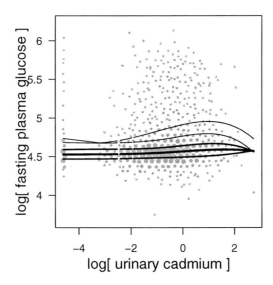

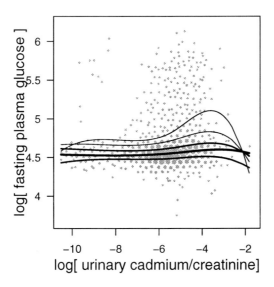

Figure 6.6   Quantile smooths (10%, 25%, 50%, 75%, 90%, 95%) of fasting plasma glucose on urinary cadmium (upper panel) and urinary cadmium:creatinine ratio (lower panel) in NHANES III

```
hanes < -read.dta("~/SafariCrap/nhanes88tsl.dta")
d2 <- read.dta("~/SafariCrap/nhanes88tsl2.dta")
hanes <- merge(hanes, d2, by="seqn")
dhanes <- svydesign(id=~sdppsu6, strat=~sdpstra6,
   weight=~wtpfsd6, nest=TRUE,
   data=subset(hanes, !is.na(wtpfsd6)))
dhanes <- update(dhanes,
   fpg=ifelse(phpfast>=8 & g1p<8000, g1p, NA))
dhanes <- update(dhanes, diab=cut(fpg,c(0,110,125,Inf)),
   diab1=cut(fpg, c(0,100,125,Inf)))
dhanes <- update(dhanes, cadmium=ifelse(udp<88880,udp,NA),
   creatinine=ifelse(urp<88880,urp,NA))
dhanes <- update(dhanes,
   age=ifelse(hsaitmor>1000, NA, hsaitmor/12))

model0 <- svyolr(diab~cadmium+creatinine, design=dhanes)
model1 <- svyolr(diab~cadmium+creatinine+age+hssex,
   design=dhanes)
model2 <- svyolr(diab~(cadmium+creatinine)*I(age-40)+hssex,
   design=dhanes)
model2a <- svyolr(diab~(cadmium+creatinine)+age+hssex,
   design=subset(dhanes, age>55))
model2b <- svyolr(diab~(cadmium+creatinine)+age+hssex,
   design=subset(dhanes, age<=55))
model2c <- svyolr(diab~(cadmium+creatinine)+age+hssex,
   design=subset(dhanes, age<=40 & had1!=1))
```

Figure 6.7 Associations between ADA diabetes category and urinary cadmium concentration

be -1 if they were in the same units). The proportional odds model is fitted with svyolr(), and the model formula looks just as it would for logistic regression, except that the outcome variable diab is a three-level factor rather than a binary variable. The simplest model, model0, gives a coefficient of 0.326 for cadmium, with a p-value of 0.00014, so there is strong evidence of an association. Exponentiating the coefficient gives an odds ratio of 1.39.

The coefficient for creatinine is -0.0024, not far from -0.01×0.326. Fitting separate logistic regression models for the two cutpoints gives coefficients of 0.22 and 0.35, so there does not seem to be a strong objection to averaging over the two cutpoints as a proportional odds model does.

As diabetes risk increases sharply with age, age is a potentially important confounder. It is also standard practice to adjust almost every model in epidemiology for sex. After adding these variables, model1 gives a much smaller coefficient, 0.068 with a standard error of 0.05, little evidence of association. The association with diabetes may be different at different ages: not only are there two completely different types of diabetes, which occur at different ages, but even Type II (or 'adult') diabetes may have different causes at different ages.

Fitting model2 reveals a strong interaction between cadmium and age. The coefficient for cadmium, which in this model estimates the association at age 40, is 0.14, with a standard error of 0.085. The interaction term with age, which estimates the amount by which the coefficient changes for each year of age, is -0.0093. This rate of change suggests that the association will reverse after about age 55, and fitting model2a restricted to those over 55 confirms this: the coefficient is -0.088, which is small and in the reverse direction. Conversely, fitting model2b and similar models with lower age thresholds confirms that the association is stronger at younger ages.

A possible explanation for this pattern of association by age would be that the younger participants are more likely to have Type I diabetes and that this is more strongly associated with urinary cadmium, perhaps because Type I diabetes occurs earlier and has had more time to cause kidney damage. The data set does not specifically distinguish Type I and Type II diabetes, but since Type I diabetes is rarely left undiagnosed for long, excluding everyone with a clinical diagnosis of diabetes should remove most of the Type I diabetics. In fact, in the subset without a prior clinical diagnosis of diabetes (in model2c) the association is slightly stronger than in everyone under 40. The estimated coefficient is 0.42, with a standard error of 0.126, and a p-value of 0.001, so the stronger association at earlier ages is apparently not due to Type I diabetes.

Converting the coefficient to an odds ratio gives 1.52, per one unit difference in cadmium. A one-unit difference in urinary cadmium is a difference of 100 ng/l. It is easier to interpret this if we examine the distribution of urinary cadmium. The interquartile range is 0.7 for the whole sample, or 0.5 for the subset used in model2c, so it is reasonable to consider a difference of 0.5 units in urinary cadmium concentration. Using svycontrast(), the odds ratio for a 0.5 unit difference in cadmium is $\exp(0.42*0.5)$, or 1.23. A 95% confidence interval is $(1.09, 1.40)$. It is entirely possible that confounding by other unmeasured variables is responsible for this association. An obvious possibility is geographic variation in both diabetes risk and environmental cadmium concentrations. If the association is due to a real effect of cadmium, it is puzzling that it is stronger at earlier ages.

The association at earlier ages is substantially weakened when the sampling weights are not used (e.g., an unweighted version of model2c gives a coefficient of only 0.1 for cadmium). As suggested in section 5.3, this could be due to a few influential points with high weight. Removing the 35 points with the highest weight (about 1% of the data) gives a slight increase in the coefficient, and removing the 150 points with the highest weight gives only a small decrease. In this example the difference between the weighted and unweighted analyses does not seem to be due to a few influential points, so there appears to be a systematic bias in the unweighted analysis. Differences between regions of the country or between urban and rural areas might explain the bias.

6.2.1   Other cumulative link models

The logit function in equation 6.2 can be replaced by other *link functions*. For example, the complementary log-log link

$$\log(-\log \Pr[Y > k]) = \alpha_k + x\beta. \tag{6.3}$$

gives a discrete analogue of the proportional hazards model for survival. Other link functions supported by svyolr() include the probit, which is the inverse of the Normal cumulative distribution function, and the cauchit, which is influenced less by the response at extreme values of X. These models are collectively called *cumulative link models*.

There are alternative ways to write the parameters in these models. In equations 6.2 and 6.3, α_1 is used to compute $\Pr[Y > 1]$ and α_2 is used to compute $\Pr[Y > 2]$. Since $\Pr[Y > 2]$ is smaller than $\Pr[Y > 1]$, $\alpha_2 < \alpha_1$. Also, if higher values of x are associated with higher values of Y then equations 6.2 and 6.3 imply that $\beta > 0$. Some researchers define α_k to have the opposite sign from equations 6.2 and 6.3, so that $\alpha_2 > \alpha_1$, and the model is occasionally written so that β is negative if higher values of x are associated with higher values of Y. When reading a paper or using unfamiliar software it is important to verify which notation is being used.

It is also possible to construct these cumulative link models as linear regression models for an unobserved continuous response variable that is then chopped up into the observed categories. For example, the ordinal probit model

$$\Phi^{-1}(\Pr[Y > k]) = \alpha_k + x\beta \tag{6.4}$$

where $\Phi()$ is the Normal cumulative distribution function, can be viewed as a linear regression model for an unobserved variable Z

$$Z = \tilde{\alpha} + x\beta + \epsilon \tag{6.5}$$

where ϵ have a Normal distribution. The values α_k then describe how Z is divided into categories to give the discrete observed variable, Y.

It is not necessary to believe in Z to use an ordinal model, and it can be dangerous to regard the usefulness of a proportional odds or probit model as evidence for the reality of an underlying latent variable. This *reification problem* is discussed at great length in the literature on intelligence testing and factor analysis [170, 171].

6.3 LOGLINEAR MODELS

Loglinear models are a class of multivariate models for categorical data. The term "multivariate" means that loglinear models can treat the relationships among set of variables symmetrically, without specifying which are predictors and which is the response. The tests of independence in contingency tables that were discussed in section 2.4.2 are very simple comparisons of loglinear models.

The code in Figure 2.9 computed a two-way table of health insurance status (Yes/No) by smoking status (Current/Former/Never) in the California Health Interview Survey. If we write p_{ij} for the estimated population probability of being in smoking category i and insurance category j, the Rao–Scott tests of independence performed using

```
svychisq(~smoking+ins, chis)
```

is comparing a loglinear model where smoking and insurance are independent to a loglinear model allowing arbitrary dependence. The formulas for these models are

$$\log p_{ij} = \log p_0 + \beta_i + \gamma_j \tag{6.6}$$

to the model

$$\log p_{ij} = \log p_0 + \beta_i + \gamma_j + \kappa_{ij} \tag{6.7}$$

where each Greek letter represents a set of scores that add up to zero and are uncorrelated with the other scores. Equation 6.6 says that separate scores for smoking and insurance add to give the log-probability. Exponentiating this shows that the probability of being in smoking category i and insurance category j is obtained by multiplying the probability of being in smoking category i and the probability of being in insurance category j. That is, smoking and insurance status are independent.

An independence model like equation 6.6, with a separate score for each variable, is the simplest reasonable model for a multiway table. A model without γ_j, for example, would say that the cell probability does not depend on insurance status, that is, the probability of being insured is the same as the probability of being uninsured. Equation 6.7, called a *saturated* model is at the other extreme. It places no constraints on the data, so that the modeled probabilities p_{ij} will be just the observed probabilities for each cell of the table.

Figure 6.8 shows how to fit these two models using the `svyloglin()` function and compares the results to those from `svychisq()`. It is first necessary to drop the empty factor levels (corresponding to missing data before imputation), as otherwise these levels would appear in the two-way table. This is done in the first two lines of code.

The call to `svyloglin()` for the independence model uses a model formula and a survey design object in much the same way as other survey analysis functions. The call for the saturated model uses the `update()` function to add terms to the independence model. The dot in the model formula represents the model terms from the existing model, `null`, and an additional interaction term is added with scores that depend on both smoking and insurance status (the κ_{ij} terms in equation 6.7). Using `update()` to add terms to an existing model is much faster than using `svyloglin()` because most of the computational effort in `svyloglin()` comes in the initial setup of the multiway table, which is the same for all the models.

The `anova()` function compares two nested loglinear models to assess the evidence for the additional terms. At the time of writing, this function is rather picky about its inputs: the two models to be compared must be based on exactly the same multiway table, meaning that the formulas must have specified the same variables in the same order. Using `update()` is a simple way to ensure the models are compatible in this way. As for `svychisq()`, there are several testing options. The test can be based on the score (Pearson χ^2) statistic or on the deviance difference between the models, as described by Rao and Scott[130]. The default is to give both test statistics. There are also four possible approximations to the distribution of the test statistic. The default is the second-order Rao–Scott correction; the other options are described in Appendix A.3. The output is different from that of `svychisq()`, but the p-value for the default version of `svychisq()` is the same as for the score statistic in `anova()`, as shown by the explicit calculation with the F distribution function `pf()`. To see that the test statistic is also the same it is necessary to ask `svychisq()` to report the unadjusted statistic, which is done with the second call.

The `summary()` methods for these loglinear models give the estimates and standard errors for all the coefficients except the intercept term ($\log p_0$ in equations 6.6 and 6.7). Some coefficients are redundant and are not explicitly given. For example, of the six κ_{ij} terms, only two are needed. These two, combined with the sum-to-zero constraints on scores, imply the values of the remaining coefficients. By calling `model.matrix()` on the model object it is possible to match the coefficients to contrasts between factor levels explicitly. The large positive coefficient for the first interaction term means that current smokers are more likely than former smokers to be uninsured (perhaps because they are younger). The second interaction term is small, indicating that there is not much difference between former smokers and never-smokers in insurance status.

6.3.1 Choosing models.

With more than two variables there is a greater number of potentially interesting models to consider. This section will provide only a brief overview of model choice; for more detail consult references such as Bishop, Fienberg & Holland[10] or Christensen [34]. The models are still of the general form given in equation 6.7, with scores depending on a single variable (such as β_i, γ_j) or on combinations of two or more variables (such as κ_{ij}). Combination (interaction) terms are specified with the

```
> droplevels<-function(f) as.factor(as.character(f))
> chis<-update(chis, smoking=droplevels(smoking),
    ins=droplevels(ins))
> null <- svyloglin(~smoking+ins, chis)
> saturated <- update(null, ~.+smoking:ins)
> anova(null, saturated)
Analysis of Deviance Table
 Model 1: y ~ smoking + ins
Model 2: y ~ smoking + ins + smoking:ins
Deviance= 659.1779 p= 6.638917e-33
Score= 694.3208 p= 5.352784e-34
> svychisq(~smoking+ins, chis)
Pearson's X^2: Rao & Scott adjustment

data:  svychisq(~smoking + ins, chis)
F = 130.1143, ndf = 1.992, ddf = 157.388, p-value < 2.2e-16
> pf(130.1143,1.992,157.388,lower.tail=FALSE)
[1] 5.327203e-34
> svychisq(~smoking+ins, chis, statistic="Chisq")
Pearson's X^2: Rao & Scott adjustment

data:  svychisq(~smoking + ins, chis, statistic = "Chisq")
X-squared = 694.3208, df = 2, p-value < 2.2e-16
> summary(null)
Loglinear model: svyloglin(~smoking + ins, chis)
                coef         se p
smoking1 -0.6209497 0.012511031 0
smoking2  0.7600805 0.008386853 0
ins1     -0.8246480 0.010158262 0
> summary(saturated)
Loglinear model: update(null, ~. + smoking:ins)
                    coef         se            p
smoking1      -0.4440871 0.01612925 7.065991e-167
smoking2       0.7527001 0.01295493  0.000000e+00
ins1          -0.8052182 0.01151648  0.000000e+00
smoking1:ins1  0.2821710 0.01847008  1.085166e-52
smoking2:ins1 -0.0310743 0.01482575  3.608497e-02
> model.matrix(saturated)
 (Intercept) smoking1 smoking2 ins1 smoking1:ins1 smoking2:ins1
1           1        1        0    1             1             0
2           1        0        1    1             0             1
3           1       -1       -1    1            -1            -1
4           1        1        0   -1            -1             0
5           1        0        1   -1             0            -1
6           1       -1       -1   -1             1             1
```

Figure 6.8 Testing independence of insurance status and smoking in CHIS

: notation, and using a * instead of a : requests all simpler terms as well, so that smoking*ins is the same as ~smoking+ins+smoking:ins.

One important subset of models is the *graphical models*. These are models that describe *conditional independence* relationships between variables. For example, the relationship between smoking and health insurance might be explained by age, so that within each age group there is no association between smoking and insurance. In this scenario, smoking and insurance are conditionally independent given age. A loglinear model with smoking:age and insurance:age interaction terms represents this conditional independence, and could be compared to the saturated model to test the hypothesis of conditional independence.

The name *graphical model* comes from the representation of conditional independence relationships in a mathematical graph or network (e,g., Figure 6.11). A line connects two variables if their relationship is explicitly modelled, with no line if the two variables are conditionally independent given the modelled relationships. The lines in the graph specify all the two-variable interaction terms in the model, and interaction terms with three or more variables are included when all the two-variable terms for pairs of those variables are in the model.

A broader class is the *hierarchical* loglinear models.. These models specify subtables of the full multiway table of variables where the observed counts must match the fitted counts. In equation 6.6, the fitted proportions in one-way tables for smoking status and insurance status must match the observed proportions. In equation 6.7, the full two-way table of fitted proportions must match the observed proportions. In a hierarchical loglinear model whenever an interaction term with three or more variables is included, all interaction terms of subsets of those variables must also be included. This will happen automatically in svyloglin() if the * notation is used to specify interaction terms.

Example: neck and back pain in NHIS. Each year, the National Health Interview Survey (NHIS) interviews a probability sample of the adult non-institutionalized civilian population of the US. The 2006 survey sampled 428 geographically defined PSUs from a population of approximately 1900. Within each PSU, housing units that existed at the previous census are sampled in small geographical clusters, and building permits for housing units since the previous census are sampled separately, again in small clusters. Households with black, Asian, or Hispanic members are oversampled. One adult for each sampled household is interviewed about their own health and that of household members. The 2006 "sample adult" file contains data on 552 variables for 24725 interviewed adults. We will examine data on neck pain and lower back pain.

The National Center for Health Statistics provides SAS, SPSS, and Stata command files for loading the data. I used the Stata command files to create a Stata binary file, read this into R with read.dta(), and then stored it in a SQLite database. Figure 6.9 gives the code for creating a survey design object from the database and defining some new variables. The variable sickleave indicates whether the participant's job provides paid sick leave, the variables neckpain and backpain indicate pain in the neck or in the lower back during the past three months.

```
nhis<-svydesign(id=~psu_p,strat=~strat_p,data="nhis",
  dbtype="SQLite", dbname="nhis.db",nest=TRUE)

nhis<-update(nhis, agegp=cut(age_p,c(0,30,50,65,Inf)),
  bmigp=cut(bmi,c(0,25,30,35,Inf)))
nhis<-update(nhis,
  sickleave=factor(ifelse(pdsicka=="1 Yes","yes","no")))
nhis<-update(nhis,
  backpain=factor(ifelse(painlb=="1 Yes","yes","no")))
nhis<-update(nhis,
  neckpain=factor(ifelse(paineck=="1 Yes","yes","no")))
```

Figure 6.9   Defining a survey design object for the National Health Interview Survey, 2006

Figure 6.10 constructs six loglinear models for the four-way table of sickleave, sex, backpain, and neckpain. The independence model for this four-way table is model a. The deviance for this model, a measure of goodness of fit, is obtained with deviance(a), which gives 3575.8. With simple random sampling the deviance for a correctly specified model has expected value equal to the number of degrees of freedom in the model, that is, the number of cells in the table minus the number of parameters. In a complex design the relationship is not that simple, but the ratio of deviance to degrees of freedom is related to the design effects and is unlikely to be more than 2–3 for a correctly specified model. The degf() function gives the number of degrees of freedom, which is 11 for model a, much smaller than the deviance.

Model a2 includes all two-variable interactions and model a3 includes all three-variable interactions. These have deviance of 12.0 and 0.4 with 5 and 1 degrees of freedom, respectively, so it is not immediately obvious whether they fit well. Using the anova() function to compare model a to a2 gives a deviance difference of 3564 and a p-value that is effectively zero, so a2 fits much better than a. A similar comparison of models a2 and a3 gives a deviance difference of 11.5, and a p-value of 0.02 from the Rao–Scott test. Model a3 fits better than model a2, but a p-value of 0.02 from a data set of this size is not particularly impressive.

Model a2 is a reasonable starting point for further model building. Models b1, b2, and b3 examine whether an interaction is needed between neck pain and back pain, and whether interactions with sickleave are needed. Model b1 fits much better than b2, with a deviance difference of 3326 and a p-value of effectively zero, indicating that neck pain and back pain do go together. Models b1 and b3 differ in that b1 has the three-variable interaction between backpain, neckpain and sex and b3 has only the three two-variable interactions among these three variables. The deviance difference reported by anova(b1, b3) is 1.9 with a p-value of 0.16, so the models do not differ importantly in their fit.

Comparing model b3 to a2 tests whether an interaction between paid sickleave and back or neck pain is important, since these are the two two-variable interaction

```
a<-svyloglin(~backpain+neckpain+sex+sickleave, nhis)
a2<-update(a, ~.^2)
a3<-update(a,~.^3)
anova(a, a2)
anova(a2, a3)
b1<-update(a,~.+(backpain*neckpain*sex)+sex*sickleave)
b2<-update(a,~.+(backpain+neckpain)*sex+sex*sickleave)
b3<-update(a,~.+backpain:neckpain+sex:backpain+sex:neckpain
    +sex:sickleave)
```

Figure 6.10   Models for back pain and neck pain in NHIS 2006

terms not included in b3. The deviance difference is 49, with a p-value of 2×10^{-9}, so there is strong evidence for a relationship. Using summary(a2) to extract the coefficients and standard errors shows that the coefficients for both interaction terms are negative: people in jobs with paid sick leave are less likely to report back or neck pain. The coefficient are relatively small, -0.04 for back pain and -0.01 for neck pain. Exponentiating these coefficients gives odds ratios of 0.96 and 0.99, so the difference in reporting pain is only a few percent. This contrasts with the coefficient of 0.55 for the interaction between neck pain and back pain, which gives an odds ratio of $e^{0.55} = 1.73$.

Both b1 and b2 are graphical models, with the conditional independence graphs shown in Figure 6.11, but b3 is not a graphical model, since it has all three two-variable interactions for backpain, neckpain, and sex without the three-variable interaction. Model b3 thus does not have an interpretation in terms of conditional independence.

All six models are hierarchical loglinear models. Model a matches fitted and observed proportions for the four one-way tables. Model a2 matches fitted and observed proportions for the six two-way tables possible for four variables. Model b1

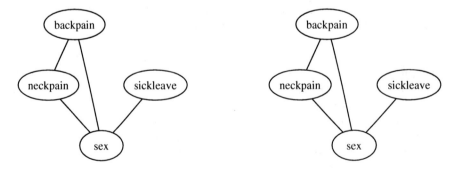

Figure 6.11   Graphical models for back pain and neck pain in NHIS 2006. Model b2 is on the left, model b1 is on the right.

matches fitted and observed proportions for the three-way table of `sex`, `backpain`, and `neckpain`, and for the two-way table of `sex` and `sickleave`.

The conclusion from this analysis would be that there is a strong association between reporting neck pain and reporting back pain, and that people in jobs with paid sickleave are slightly less likely to report recent neck pain or lower back pain. Many more variables might be included in a more thorough analysis of these data. Age is likely to be important, and the NHIS files include occupation category and industrial category for jobs, which are also clearly relevant.

6.3.2   Linear association models

In the Scottish Household Survey analysis in section 6.1, internet use was reduced to a binary "Yes"/"No" even though the SHS collected data on the amount of internet use. The variable `rc5` has five levels ranging from "up to 1 hour/week" to "more than 20 hours/week". A set of loglinear models could be fitted to examine the relationships between age group, income, and amount of internet use, but the models described so far take no account of the ordering of categories in these variables. In the analysis of neck and back pain this did not matter, since all the variables we used were binary.

One approach to incorporate the ordering of categories in a loglinear model is to assign numeric codes to the categories and then reduce the complete set of interaction scores (κ_{ij} in equation 6.7) to a single score obtained by multiplying these numeric codes. Different sets of numeric codes give rise to different models, but it is often sufficient simply to use 1, 2 3, ... as codes. Using these codes the coefficient for the interaction term can be interpreted as a log odds ratio in a 2×2 subtable of adjacent cells in the two-way table.

Figures 6.12 and 6.13 show an analysis of the amount of internet use and its relationship to age and to income. There are three models for age and three models for income. The first is the independence model. This is the same as for unordered categories, since it describes only the proportions in each category in a single-variable table, and the ordering of categories has no special implications for the numbers in each category. The second model is the linear association model, which uses numeric codes 1, 2, 3, ... obtained with `as.numeric()`. The third is the saturated model, as in equation 6.7. Comparing the linear association model and the independence model shows whether the linear scores capture a substantial association between the variables. Comparing the linear association model and the saturated model shows whether there is a substantial component of association not captured by the linear scores.

The models for income in Figure 6.12 show a fairly strong association with amount of internet use in the saturated model, though much weaker than the association with any use vs no use estimated in Figures 6.2 and 6.5. The linear association model captures about 40% of the deviance difference between the independence and saturated models in a single association parameter, in contrast to the 16 association parameters of the saturated model. The difference between the linear association model and the saturated model is not statistically significant: the remaining deviance

```
> shs <- update(shs,
    income=ifelse(groupinc=="missing", NA, groupinc))
> null <- svyloglin(~rc5+income, shs)
> saturated <- update(null, ~.^2)
> anova(null, saturated)
Analysis of Deviance Table
 Model 1: y ~ rc5 + income
Model 2: y ~ rc5 + income + rc5:income
Deviance= 289.2165 p= 4.950754e-08
Score= 282.5453 p= 9.021286e-08
> lin <- update(null, ~.+as.numeric(rc5):as.numeric(income))
> anova(null, lin)
Analysis of Deviance Table
 Model 1: y ~ rc5 + income
Model 2: y ~ rc5 + income + as.numeric(rc5):as.numeric(income)
Deviance= 105.3609 p= 7.919452e-07
Score= 105.6157 p= 7.681001e-07
> anova(lin, saturated)
Analysis of Deviance Table
 Model 1: y ~ rc5 + income + as.numeric(rc5):as.numeric(income)
Model 2: y ~ rc5 + income + rc5:income
Deviance= 183.8556 p= 0.09605066
Score= 184.0606 p= 0.09591673
```

Figure 6.12   Linear association tests for amount of internet use in Scotland, data from Scottish Household Survey

```
> shs <- update(shs, agegp=cut(age, c(20, 40, 60, 100)))
> null <- svyloglin(~agegp+rc5,shs)
> sat <- update(null, ~.^2)
> lin <- update(null, ~.+as.numeric(rc5):as.numeric(agegp))
> anova(null,lin)
Analysis of Deviance Table
 Model 1: y ~ agegp + rc5
Model 2: y ~ agegp + rc5 + as.numeric(rc5):as.numeric(agegp)
Deviance= 350.1315 p= 3.038603e-18
Score= 340.5920 p= 8.628168e-18
> anova(lin,sat)
Analysis of Deviance Table
 Model 1: y ~ agegp + rc5 + as.numeric(rc5):as.numeric(agegp)
Model 2: y ~ agegp + rc5 + agegp:rc5
Deviance= 45.60211 p= 0.1813459
Score= 48.02927 p= 0.1615965
```

Figure 6.13   Linear association tests for amount of internet use in Scotland, data from Scottish Household Survey

```
null <- svyloglin(~agegp + income + rc5, shs)
m1 <- update(null, ~.+as.numeric(agegp):as.numeric(rc5))
m2 <- update(m1, ~. +agegp:income)
m3 <- update(m2, ~.+as.numeric(income):as.numeric(rc5))
m4 <- update(m2, ~.+income:rc5)
full <- update(null, ~.^2)
```

Figure 6.14   Modelling amount of internet use in Scotland, from Scottish Household Survey

difference is not much larger than would be expected by chance when fitting 15 extra parameters.

The models for age group in Figure 6.13 show the same patterns, even more strongly. The deviance difference between the linear association model and the null model is 350, for one parameter, and the difference between the linear association model and the saturated model is only 45.6, for 11 extra parameters. The linear association model captures the association between amount of internet use and age quite accurately. The coefficient for the linear association term is -0.19, corresponding to an odds ratio of $e^{-0.19} = 0.83$ for a 2×2 subtable of adjacent cells in the 5×4 table of amount of internet use by age group.

A starting point in constructing a model for all three variables would be a model with linear associations for agegp by rc5 and income by rc5. The code is in Figure 6.14. A linear association between income and age does not appear plausible, and the full interaction for this term seems appropriate. It is not possible to compare

Table 6.1 Comparisons of models of internet use in Scotland. Each model is compared to the next.

Model	Deviance	Deg. Freedom	Added Term	p-value
null	1934	64	linear age:internet	10^{-17}
m1	1603	63	full age:income	10^{-50}
m2	481	55	linear income:internet	10^{-7}
m3	373	54	full income:internet	0.001
m4	188	39	full age:internet	0.263
full	150	32		

to the saturated model, since the full three-way table has zero cells and the Rao–Scott formulas do not straightforwardly apply. Table 6.1 shows the deviance and degrees of freedom for each of these models.

The ratio of deviance to degrees of freedom for model `full` is 4.7. Under simple random sampling this ratio should be close to 1.0 and a value of 4.7 would indicate a serious lack of fit. The option of a simple test comparing this model to a saturated model is unavailable, but some guidance can be obtained by considering the scale factors (or "generalized design effects") used in testing m4 against `full`. These scale factors are returned in the `$a` component of `anova(m4, full)` and range from 3.6 to 5.4, with a mean of 4.3. A ratio of 4.7 is well within this range, so to the extent that these scale factors would be the same in comparing `full` to a saturated model they suggest the model fits reasonably well.

EXERCISES

6.1 Using the same data as in Section 5.2.4, define hypertension as systolic blood pressure greater than 140 mmHg or diastolic blood pressure greater than 90 mmHg. Fit logistic regression models to investigate the association between dietary sodium and potassium and hypertension.

6.2 This exercise uses the Washington State Crime data for 2004 as the population. The data consist of crime rates and population size for the police districts (in cities/towns) and sheriffs' offices (in unincorporated areas), grouped by county.

    a) Take a simple random sample of 10 counties from the state and use all the data from the sampled counties. Estimate the total number of murders and of burglaries in the state.

    b) Fit a Poisson regression (`family=quasipoisson`) to model the relationship between number of murders and population. Poisson regression fits a linear model for the logarithm of the mean of the outcome variable. If the murder rate were constant, the optimal transformation of the predictor variable would be the logarithm of population, and its coefficient would be 1.0. Is this supported by the data?

c) Predict the total number of murders in the state using the Poisson regression model. Compare to a ratio estimator using population as the auxiliary variable.

d) Take simple random samples of five police districts from King County and five counties from the rest of the state. Fit a Poisson regression model with population and stratum (King County vs elsewhere) as predictors. Predict the total number of murders in the state.

6.3 The variable MISEFFRT asks "How often in the past 30 days did you feel that everything was an effort?", on a 1–5 scale with 1 meaning "All" and 5 meaning "None". Investigate whether this variable varies seasonally and whether it peaks in winter by defining predictor variables cos(IMONTH * 0.5236) and sin(IMONTH * 0.5236), which describe smooth annual cycles, and fitting:

a) a logistic regression model with outcome MISEFFRT = 5
b) a linear regression model with outcome MISEFFRT
c) What further modeling could you do to investigate whether sunlight intensity was related to this seasonal variation.

6.4 Using the California Health Interview Study 2005 data,

a) fit a logistic regression model to the relationship between the probability of having health insurance (ins) and household annual income (ak22_p). Important potential confounders include age (srage_p), sex (srsex) and race (racecen), and interactions may also be important.
b) What would be the impact on the interpretation of the income coefficient of adding a variable to the model indicating whether an employer provided health insurance?

6.5 Using the California Health Interview Study 2005 data, fit a relative risk regression model to the relationship between the probability of having health insurance (ins) and household annual income (ak22_p), age (srage_p), sex (srsex) and race (racecen). Compare the coefficients to those from a logistic regression.

6.6 Using the same data as in Section 5.2.4, create an ordinal blood pressure variable based on systolic and diastolic pressure with categories "normal" (systolic < 120 and diastolic < 80), "prehypertension" (higher than normal but systolic < 140 and diastolic < 90), "hypertension stage 1" (higher than prehypertension but systolic < 160 and diastolic < 100), and "hypertension stage 2" (systolic at least 160 or diastolic at least 100). Fit proportional odds regression models to investigate the association between dietary sodium and potassium and hypertension.

6.7 Using data for Florida (X_STATE =12) from the 2007 BRFSS, fit a loglinear model to assocations between the following risk factors and behaviors: perform vigorous physical activity (VIGPACT), eat five or more servings of fruit or vegtables per day (X_FV5SRV), binge drinking of alcohol (X_RFBING4), ever had an HIV test (HIVST5), ever had Hepatitis B vaccine (HEPBVAC), age group (X_AGE_G), and sex (X_SEXG). All except age group are binary. You will need to remove missing values (coded 7, 8, or 9).

CHAPTER 7

POST-STRATIFICATION, RAKING, AND CALIBRATION

In which the whole tells us about some of the parts.

7.1 INTRODUCTION

Chapter 2 showed the increase in precision that can come from using population data to stratify sampling. Stratification is not always a desirable way to use these population data: there may be too many potential stratification variables, the best strata may be different for different analyses, or the need for cluster sampling may prevent stratification on individual-level variables. Population data may also be available in a form that does not allow for stratification — random-digit dialing, for example, cannot easily stratify on any individual characteristics, because these characteristics are not known before dialing and so cannot be used to select the sample.

This chapter deals with techniques for using known population totals for a set of variables (*auxiliary variables*) to adjust the sampling weights and improve estimation for another set of variables. All of these techniques have the same idea: adjustments

Complex Surveys: A Guide to Analysis Using R. By Thomas Lumley
Copyright © 2010 John Wiley & Sons, Inc.

are made to the sampling weights so that estimated population totals for the auxiliary variables match the known population totals, making the sample more representative of the population. A second benefit is that the estimates are forced to be consistent with the population data, improving their credibility with people who may not understand the sampling process.

There are two quite different applications of these techniques. One is to increase precision of estimation, the other is to reduce the bias from nonresponse, especially *unit non-response*, where a sampled individual refuses to participate or otherwise provides no information for analysis. From the computational viewpoint the main difference between these applications is in the criteria for choosing auxiliary variables, which are discussed in section 7.6. The use for non-response is the more important of the two in large-scale surveys, but it has largely been completed by the time the data arrive at the typical user. Calibration to improve precision will be important in the two-phase epidemiologic designs discussed in Chapter 8. *Item non-response*, where some variables but not others were observed for an individual is addressed in Chapter 9.

7.2   POST-STRATIFICATION

The simplest technique for adjusting sampling weights is *post-stratification*. Suppose we have a division of the population into groups. One possible design would be a stratified random sample using these groups as strata, sampling n_k individuals from population stratum k containing N_k individuals. The sampling weights are $1/\pi_i = N_k/n_k$ and the Horvitz–Thompson estimator $\hat{N}_k$ of the population group size is exactly correct and its standard error is zero. Estimation of other population totals is improved by removing the variability between groups: the contribution of each group to a population total is fixed by design, rather than random.

If N_k were known, but the sampling was not stratified, the estimated population group sizes would not be exactly correct. Post-stratification adjusts the sampling weights so that the estimated population group sizes are correct, as they would be in stratified sampling. The sampling weights $1/\pi_i$ are replaced by weights g_i/π_i where $g_i = N_k/\hat{N}_k$ for the group containing individual i. The estimated group size for the kth group will then be

$$n_k \times \frac{g_i}{\pi_i} = n_k \times \frac{1}{\pi_i} \times \frac{N_k}{\hat{N}_k} = \hat{N}_k \times \frac{N_k}{\hat{N}_k} = N_k.$$

The estimated group size is always the same as the actual group size, and just as for a stratified sample the contribution of each group to population totals is now fixed. These groups are often called *post-strata*.

One issue has been glossed over here. If we are unlucky enough to sample no-one from group k it is not possible to perform the re-weighting. As long as the expected sample sizes in each group are not too small, the probability of ending up with no observations in a group is very small. For example, if the expected sample size in the group is only 10, the probability of having at least one observation in the group is

99.995%. On the rare occasions when there are no observations in the group it would be necessary to merge two post-strata or make some other *ad hoc* correction. If we approximate the standard errors by neglecting the very rare occasions when a group ends up empty, post-stratification on the group variable does give the same standard errors as sampling stratified on the group variable with the same group sample sizes.

The same correction can be used for any sampling design, not just for simple random sampling. If the totals $\{N_k\}$ for a set of population groups are known and the Horvitz–Thompson estimates $\{\hat{N}_k\}$ are computed, the adjustment g_i to the sampling weight for individual i is $N_k/\hat{N}_k$ for the group k containing the individual. As the sample size increases, $\hat{N}_k$ will become more accurate and so the adjustment $g_i = N_k/\hat{N}_k$ will become closer and closer to 1, but the proportional reduction in standard errors from post-stratification will remain roughly constant.

Adjusting the weights to incorporate the auxiliary information is straightforward, but estimating the resulting standard errors is more difficult. For replicate-weight designs it is sufficient to post-stratify each set of replicate weights (Valliant [178]). Since the estimated group size will then match the known population group size for every set of replicates, between-group variation will not produce differences between replicates and so the replicate-weight standard errors will correctly omit the between-group differences.

Standard errors for a population total are computed by decomposing the variable into stratum means and residuals. Writing μ_k for the true population mean in group k and $\mu_{k(i)}$ for the mean in the group containing observation i, $Y_i = (Y_i - \mu_{k(i)}) + \mu_{k(i)}$. The variance of the estimated total of Y is

$$\text{var}\left[\hat{T}_y\right] = \text{var}\left[\sum_{i=1}^{n}\frac{1}{\pi_i}\left(Y_i - \mu_{k(i)}\right)\right] + \text{var}\left[\sum_{i=1}^{n}\hat{N}_{k(i)}\frac{\mu_{k(i)}}{\pi_i}\right] \qquad (7.1)$$

and because $\hat{N}_k$ is fixed, the second variance is zero. This argues for estimating the variance of a post-stratified estimator by subtracting off the group means and taking the Horvitz–Thompson estimator of the variance of the residuals. The variance estimator will not be unbiased, since the residuals from the estimated group mean will tend to be smaller than the residuals from the true group mean. When post-stratifying a simple random sample this bias could be computed and corrected by multiplying group k's contribution to the variance by $n_k/(n_k - 1)$, but there is no simple correction that works for cluster-sampled or multistage designs. The bias in the variance estimator can safely be ignored as long as the group sizes n_k are not too small. The same procedure extends to summary statistics that solve a population equation: the contributions to this population equation are centered at the group means before computing their variances [133].

Two issues this leaves open are whether the group means should be estimated using $1/\pi_i$ or g_i/π_i as weights, and whether $1/\pi_i$ or g_i/π_i should be used as weights when computing the Horvitz–Thompson variance estimator. In the absence of non-response this choice makes very little difference, but in the presence of substantial non-response it does matter (Kott [84]). The survey package uses $1/\pi_i$ for the stratum means and g_i/π_i in the Horvitz–Thompson estimator.

```
> data(api)
> clus2_design <- svydesign(id=~dnum+snum, fpc=~fpc1+fpc2,
     data=apiclus2)
> pop.types <- data.frame(stype=c("E","H","M"),
     Freq=c(4421,755,1018))
> ps_design <- postStratify(clus2_design, strata=~stype,
     population=pop.types)
> svytotal(~enroll, clus2_design, na.rm=TRUE)
         total    SE
enroll 2639273 799638
> svytotal(~enroll, ps_design, na.rm=TRUE)
         total    SE
enroll 3074076 292584
> svymean(~api00, clus2_design)
        mean    SE
api00 670.81 30.099
> svymean(~api00, ps_design)
       mean    SE
api00   673 28.832
```

Figure 7.1   Post-stratifying a two-stage sample of schools on school type

The function `postStratify()` creates a post-stratified survey design object. In addition to adjusting the sampling weights, it adds information to allow the standard errors to be adjusted. For a replicate-weight design this involves computing a post-stratification on each set of replicate weights; for linearization estimators it involves storing the group identifier and weight information so that between-group contributions to variance can be removed.

Example: Post-stratifying on school type.   In section 3.2 we examined a two-stage sample drawn from the API population, with 40 school districts sampled from California and then up to five schools sampled from each district. Post-stratifying this design on school type illustrates the situations where improvements in precision are and are not available. The code and output are in Figure 7.1.

The first step is to set up the information about population group sizes. This information can be in a data frame (as here) or in a `table` object as is produced by the `table()` function. In the data frame format, one or more columns give the values of the grouping variables and the final column, which must be named `Freq`, gives the population counts. The call to `postStratify()` specifies the grouping variables as a model formula in the `strata` argument, and gives the table of population counts as the `population` argument. The function returns a new, post-stratified survey design object.

Post-stratification on school type dramatically reduces the variance when estimating total school enrollment across California. The standard error was approximately

800,000 for the two-stage sample and less than 300,000 for the post-stratified sample. A large reduction in variance is possible because elementary schools are about half the size of middle schools and about one-third the size of high schools on average. Post-stratification removes the component of variance that is due to differences between school types, and this component is large.

When estimating the mean Academic Performance Index there is little gain from post-stratification. A standardized assessment of school performance should not vary systematically by level of school or it would be difficult to interpret differences in scores. Since API has a similar mean in each school type, the between-level component of the variance is small and post-stratification provides little help. If this were a real survey, non-response might differ between school type, and post-stratification would then be useful in reducing non-response bias.

7.3   RAKING

Post-stratification using more than one variable requires the groups to be constructed as a complete cross-classification of the variables. This may be undesirable: the cross-classification could result in so many groups that there is a risk of some of them not being sampled, or the population totals may be available for each variable separately, but not when cross-classified.

Raking allows multiple grouping variables to be used without constructing a complete cross-classification. The process involves post-stratifying on each set of variables in turn, and repeating this process until the weights stop changing. The name arises from the image of raking a garden bed alternately in each direction to smooth out the soil. Raking can also be applied when partial cross-classifications of the variables are available. For example, a sample could be alternately post-stratified using a two-way table of age and income, and a two-way table of sex and income. The resulting raked sample would replicate the known population totals in both two-way tables.

The effect of matching observed and expected counts in lower-dimensional subtables is very similar to the effect of fitting a hierarchical loglinear model as described in section 6.3.1. In fact, the raking algorithm is widely used to fit loglinear models, where it is known as *iterative proportional fitting*. Raking could be considered a form of post-stratification where a loglinear model is used to smooth out the sample and population tables before the weights are adjusted.

Standard error computations after raking are based on the iterative post-stratification algorithm. For standard errors based on replicate weights, raking each set of replicate weights ensures that the auxiliary information is correctly incorporated into between-replicate differences. For standard errors based on linearization, iteratively subtracting off stratum means in each direction gives residuals that incorporate the auxiliary information correctly.

The rake() function performs the computations for raking by repeatedly calling postStratify(). Each of these calls to postStratify() accumulates the necessary information for standard error computations.

```
pop.ctband <- data.frame(CTBAND=1:9,
        Freq=c(515672, 547548, 351599, 291425,
                        266257, 147851, 87767, 9190, 19670))
pop.tenure <- data.frame(TENURE=1:4,
        Freq=c(1459205,493237, 128189, 156348))
frs.raked <- rake(frs.des, sample=list(~CTBAND, ~TENURE),
        population=list(pop.ctband, pop.tenure))
svymean(~HHINC, frs.raked)
svymean(~HHINC, subset(frs.raked, DEPCHLDH>0))
svymean(~HHINC, subset(frs.raked, DEPCHLDH>0 & ADULTH==1))
```

Figure 7.2   Raking on socioeconomic variables in the Family Resources Survey

Example: Family Resources Survey.   In an example in Chapter 5 population data on council tax band and housing tenure in Scotland were used to fit a linear regression model to household income and improve the estimates of mean weekly income. Raking gives another way to use this information, and allows for estimation of mean weekly income in subpopulations as well as for the whole population. Recall that in this example the weights have already been adjusted by raking before the data were provided, so that performing raking in R will not change the estimates, but will reduce the standard errors to the correct value.

The inputs to the `rake()` function are similar to those for `postStratify()`, except that a list of tables for each margin is specified rather than a single table. Figure 7.2 has code for raking the survey design object defined in Figure 5.7. In this case we specify the population tables as two data frames. Each data frame has a variable with the same name as a raking variable in the survey design object and a variable `Freq` with population frequencies for the levels of this variable. A list of two formulas specifies the names of the raking variables and a list of the two data frames (in the same order) specifies the population information. Estimating mean household income for all households using the raked design object gives £483 with standard error £7.5, the same as for the regression estimator and a substantial increase in precision over the standard error of £10.6 for the unraked design. Estimating the mean household income for all households with children gives £611, with a standard error of £12.5; before raking the estimated mean was the same, but the standard error was £15.6.

The improvement in precision is almost non-existent for the smaller subpopulation of single-parent households, where the standard error of estimated mean weekly income is £8.5 before raking and £8.4 after raking. This is partly because of the size of the subpopulation, and partly because the particular raking variables used here provide more detailed information about higher income levels. For example, households with exactly two adults and two children make up about the same fraction of the population and of the sample as single-parent households, but the estimated mean weekly income for these households is £714 with standard error £22.3 before raking and £714 with standard error £19.7 after raking, a larger gain in precision.

The benefit of raking decreases for smaller subpopulations because the full population information is less relevant to these smaller subgroups. In general, using population auxiliary information does not provide much extra precision in subpopulations unless the population information is specific to the subpopulations being analyzed. Similar conclusions hold for the impact of raking on estimation of regression coefficients, as is illustrated in the next section. Chapter 8 looks at some examples where there is detailed auxiliary information specific to a particular regression model and useful gains in precision are realized.

In addition to allowing estimation of mean income in subpopulations, raking also allows the auxiliary information to be used in estimating other statistics, such as quantiles with svyquantile(). In the original FRS design the estimated median weekly income was £355 with a 95% confidence interval £338–£372. After raking the 95% confidence interval is narrower: £341–£369.

7.4 GENERALIZED RAKING, GREG ESTIMATION, AND CALIBRATION

There are two related ways to view what is happening in post-stratification that allow extensions to continuous auxiliary variables and to a wide variety of other problems. As in section 7.2, post-stratification can be seen as making the smallest possible changes to the weights that result in the estimated population totals matching the known totals. This view of post-stratification leads to *calibration* estimators (Deville et al. [41], Deville & Särndal [40]). An alternative is to note that the estimated population total after post-stratification is a regression estimator, for a working regression model that uses indicator variables for each post-stratum as predictors (see Exercise 7.5). This view leads to *generalized regression* or *GREG* estimators (eg Särndal et al. [152]; Lehtonen and Veijanen[92]; Wu and Sitter[190]; Rao et al. [133])

A thorough review of calibration is given by Särndal [149], who distinguishes between "calibration thinking" and "regression thinking" in constructing estimators using auxiliary information. "Regression thinking" is useful in understanding why large increases in precision are possible — it is surprising to many biostatisticians that small changes in sampling weights can increase precision dramatically, but it is easier to understand that a good regression model can reduce the unexplained variation. "Calibration thinking" often leads to simpler formulas and simpler software implementation, since an explicit model is not needed.

Given auxiliary variables X_i whose population totals T_X are known, the regression estimator of the population total was constructed in section 5.2.2. A regression model is fitted to the sample to obtain regression coefficients β and the estimated population total of Y is the population total of the fitted values

$$\hat{T}_{\text{reg}} = \sum_{i=1}^{N} x_i \hat{\beta}_i = \left(\sum_{i=1}^{N} X_i \right) \hat{\beta} = T_X \hat{\beta}. \tag{7.2}$$

The parameter estimates $\hat{\beta}$ from weighted least squares estimation can be written as a weighted sum of the sampled Y_i, with weights depending on X_i and on the population totals of X. This implies that $T_X \hat{\beta}$ can also be written as a weighted sum of Y. That, is

$$\hat{T}_Y^{(\text{reg})} = \sum_{i=1}^{n} \frac{g_i}{\pi_i} Y_i$$

for *calibration weights* g_i that do not depend on Y. This is the *calibration estimator* of the population total.

Because the weights do not depend on Y, they would be the same for estimating the population total of any variable, and in particular of X, so

$$\hat{T}_X^{(\text{reg})} = \sum_{i=1}^{n} \frac{g_i}{\pi_i} X_i.$$

Since a regression estimate of the population total of X that uses the known population total of X will be exactly correct, $\hat{T}_X^{(\text{reg})} = T_X$, and the calibration weights must satisfy the *calibration constraints*

$$T_X = \sum_{i=1}^{n} \frac{g_i}{\pi_i} X_i. \tag{7.3}$$

The calibration weights g_i make the estimated and known population totals agree. If the Horvitz–Thompson estimate of T_X is too small, g_i will give more weight to large values of X; if the Horvitz–Thompson estimate is too large, g_i will downweight large values of X. When X and Y are correlated, the calibration weights that give exact estimation of T_X will also give improved estimation of T_Y.

The calibration constraints in equation 7.3 do not uniquely define the weights. For example, if there is only one auxiliary variable X it is always possible to satisfy equation 7.3 by making a large change to the weight for just one observation. One way to completely specify g_i is to also require that the calibrated weights are as close as possible to the sampling weights. That is, for some distance function $d(\,,\,)$, the calibration weights are chosen to make

$$\text{weight change} = \sum_{i=1}^{n} d\left(\frac{g_i}{\pi_i}, \frac{1}{\pi_i}\right)$$

as small as possible while still satisfying the calibration constraints. The regression estimator of the population total corresponds to one choice of $d(\,,\,)$ and the ratio estimator from section 5.1.3 to a different choice. Raking corresponds to the same choice of $d(\,,\,)$ as ratio estimation. In linear regression calibration, the calibration weights g_i are a linear function of the auxiliary variables; in raking calibration the calibration weights are a multiplicative function of the auxiliary variables.

Ratio, raking, and regression estimators were already in widespread use before calibration was developed as a unifying description. The free choice of $d(\,,\,)$ also allows new estimates to be constructed. The distance function can be chosen to force upper and lower bounds for g_i, for example, specifying that $g_i > 0.5$ and

$g_i < 2$, to prevent individual observations from becoming too influential and to prevent computational problems from zero or negative weights.

The regression and calibration approaches to using auxiliary data do not always lead to the same estimates, but they do agree for a wide range of situations that arise frequently. It is also worth noting that some of the situations where regression and calibration approaches differ according to Särndal[149] can be unified by considering regressions where the outcome variable is the individual contribution to a population equation. That is, "calibration thinking" can be duplicated by combining "regression thinking" with "influence function thinking", as it has been in biostatistics. Calibration with influence functions is described in section 8.5.1 and details for the example given by Estevao and Särndal are discussed in the Appendix, in section A.5.

Standard error estimates after calibration follow similar arguments to those for post-stratification. When estimating a population total, a variable Y_i can be decomposed into a true population regression value μ_i and a residual $Y_i - \mu_i$. An unbiased estimator of variance of the estimated totals would be the Horvitz–Thompson variance estimator applied to $Y_i - \mu_i$. Since μ_i is unknown, we apply the Horvitz–Thompson estimator to $Y_i - \hat{\mu}_i$ and obtain a variance estimator that is nearly unbiased as long as the number of parameters in the regression model is not too large (Särndal et al. [152]).

$$\mathrm{var}\left[\hat{T}_Y^{(\mathrm{reg})}\right] = \mathrm{var}\left[Y - \mu\right] \approx \mathrm{var}\left[Y - \hat{\mu}\right]. \tag{7.4}$$

When estimating other statistics the residuals are taken after linearization, as described in Appendix C.1. As usual, the estimation is more straightforward with replicate weights. Once the calibration procedure is applied to each set of replicate weights, the variances then automatically incorporate the auxiliary information.

7.4.1   Calibration in R

The survey package provides calibration with the `calibrate()` function. As with `postStratify()` and `rake()`, this function takes a survey design object as an argument and returns an updated design object with adjusted weights and all the necessary information to compute standard errors. The auxiliary variables are specified using a model formula, as for post-stratification, and the population totals are specified as the column sums of the population regression design matrix (predictor matrix) corresponding to the model formula.

Linear regression calibration.   Figure 7.3 uses calibration to repeat the estimates of US 2008 presidential election totals from Figure 5.8. The auxiliary variables are the votes in the 2000 election for George W. Bush and Al Gore. The `formula` argument specifies the auxiliary variables, and the `population` argument gives the population totals. In this example there are population totals for the intercept, and for the variables BUSH and GORE — the intercept is always included implicitly. In Figure 5.8 separate regression models were needed to estimate totals for Obama and McCain. Calibrating the survey design object incorporates the auxiliary information in all estimates, so the totals for both candidates can simply be estimated with `svytotal()`.

```
> cal.elect<-calibrate(srsdes, formula=~BUSH+GORE,
    population=c('(Intercept)'=3049, BUSH=data2000$BUSH,
      GORE=data2000$GORE))
> svytotal(~OBAMA+MCCAIN, cal.elect)
          total      SE
OBAMA   58233372 6577358
MCCAIN  51161922 4478711
> cal.elect2<-calibrate(srsdes, formula= ~BUSH+GORE,
    population=c('(Intercept)'=3049, BUSH=data2000$BUSH,
      GORE=data2000$GORE),
    variance=c(0,1,1))
> svytotal(~OBAMA+MCCAIN, cal.elect2)
          total      SE
OBAMA   60918323 4448316
MCCAIN  53509686 3353345
```

Figure 7.3   Calibrating US 2008 election data to 2000 totals

The second call to `calibrate()` specifies a different distance function $d(\ ,\ )$, one that is optimal when the variability in Y is proportional to the calibration variables. In this case, the vector c(0, 1, 1) specifies calibration weights that would be optimal if the variance of Y is proportional to $1 \times$ BUSH $+ 1 \times$ OBAMA, i.e., to the number of votes. This calibration gives very similar results to the regression estimators of totals using the working model with variance proportional to mean in Figure 5.8. The results would be identical if the coefficients from the regression models in Figure 5.8 were used in the `variance` argument.

Raking calibration.   Repeating the example of raking from the Family Resources Survey (Figure 7.2) using the `calibrate()` function gives the code in Figure 7.4. In a regression model the two factor variables CTBAND and TENURE would be coded as an intercept and an indicator variable for each category except the first. The population total for the intercept is the population size, and this is concatenated with the counts for each of the auxiliary variables, dropping the first category of each. The option `calfun="raking"` requests the calibration distance function that is equivalent to raking. The estimates of mean (and median) weekly income based on the calibrated design are identical to those obtained from the raked design in Figure 7.2. The advantage of raking calibration using `calibrate()` over raking using `rake()` is that `calibrate()` can use continuous auxiliary variables and `rake()` can use only discrete variables.

Logit calibration and bounded weights.   A popular calibration function in Europe gives so-called *logit calibration*. This distance function requires the user to specify upper and lower bounds on the calibration weights. If possible, the `calibrate()` will return calibration weights that satisfy the bounds. It may be impossible to satisfy both the bounds and the calibration constraints (Equation 7.3),

```
pop.size <- sum(pop.ctband$Freq)
pop.totals <- c('(Intercept)'=pop.size, pop.ctband$Freq[-1],
  pop.tenure$Freq[-1])
frs.cal <- calibrate(frs.des,
  formula=~factor(CTBAND)+factor(TENURE),
  population=pop.totals, calfun="raking")
svymean(~HHINC, frs.cal)
svymean(~HHINC, subset(frs.cal, DEPCHLDH>0))
svymean(~HHINC, subset(frs.cal, DEPCHLDH>0 & ADULTH==1))
```

Figure 7.4   Calibration (raking) on socioeconomic variables in the Family Resources Survey

in which case `calibrate()` will give an error. The `force=TRUE` option forces `calibrate()` to return a survey design in which the weights satisfy the bounds even if the calibration constraints are not met. This is useful to allow a series of simulations to be completed, and may be useful for data analysis if the calibration constraints are nearly met, which can be easily checked using `svytotal()`.

Figure 7.5 shows code for logit calibration on a cluster sample of school districts from the Academic Performance Index population. Using linear calibration with the same auxiliary variables gives calibration weights ranging from 0.4 to 1.8, with logit calibration the weights vary from slightly above the lower bound of 0.7 to slightly below the upper bound of 1.7.

Calibration using the 1999 Academic Performance Index gives a dramatic reduction in standard error for estimating the mean of the 2000 Academic Performance Index. In a regression model with Academic Performance Index as the outcome there is a substantial increase in precision for the intercept but not for the slope estimates. In this model the predictors are proportion of "English Language Learners" (`ell`), proportion of students who are new to the school (`mobility`) and proportion of teachers with only emergency qualifications (`emer`). These predictors and the outcome are quite strongly associated with the auxiliary variables, but this is not enough to give an increase in precision for comparisons across levels of the predictors. The problem of calibration to increase precision for coefficient estimates in regression models is discussed further in Chapter 8. Gains in precision are possible, but they require calibration targeted to the specific model. Calibration of large surveys carried out when the data is collected is not likely to increase precision for regression models fitted in secondary data analysis. On the other hand, the reduction in uncertainty for the intercept can (and in this example does) translate to a useful reduction in standard errors for fitted values.

Comparing calibration methods.   In the absence of non-response, the choice of calibration function makes little difference to the resulting estimated totals, with the difference decreasing as sample size increases [40]. In finite samples or with non-response it is possible for the results to differ, but they typically do not. Kalton and Flores-Cervantes [69] gave an artificial example of calibration comparing a range of

```
> clus1 <- svydesign(id=~dnum, weights=~pw, data=apiclus1,
    fpc=~fpc)
> logit_cal <- calibrate(clus1, ~stype+api99,
+    population= c( 6194, 755, 1018, 3914069),
+    calfun="logit",bounds=c(0.7,1.7))
> svymean(~api00, clus1)
        mean      SE
api00 644.17 23.542
> svymean(~api00, logit_cal)
        mean    SE
api00 665.46 3.42
> summary(svyglm(api00~ell+mobility+emer, clus1))
Call:
svyglm(api00 ~ ell + mobility + emer, clus1)
Survey design:
svydesign(id = ~dnum, weights = ~pw, data = apiclus1,
    fpc = ~fpc)
Coefficients:
            Estimate Std. Error t value Pr(>|t|)
(Intercept) 780.4595    30.0210  25.997 3.16e-11 ***
ell          -3.2979     0.4689  -7.033 2.17e-05 ***
mobility     -1.4454     0.7343  -1.968  0.07474 .
emer         -1.8142     0.4234  -4.285  0.00129 **

> summary(svyglm(api00~ell+mobility+emer, logit_cal))
Call:
svyglm(api00 ~ ell + mobility + emer, logit_cal)
Survey design:
calibrate(clus1, ~stype + api99, population = c(6194, 755, 1018,
    3914069), calfun = "logit", bounds = c(0.7, 1.7))
Coefficients:
            Estimate Std. Error t value Pr(>|t|)
(Intercept) 789.1015    17.7622  44.426 9.18e-14 ***
ell          -3.2425     0.4803  -6.751 3.15e-05 ***
mobility     -1.5140     0.6436  -2.352 0.038318 *
emer         -1.7793     0.3824  -4.653 0.000702 ***
> predict(m0, newdata=data.frame(ell=5,mobility=10,emer=10))
    link     SE
1 731.37 26.402
> predict(m1, newdata=data.frame(ell=5,mobility=10,emer=10))
    link     SE
1 739.96 15.02
```

Figure 7.5   Logit calibration on cluster sample of school districts from the Academic Performance Index population

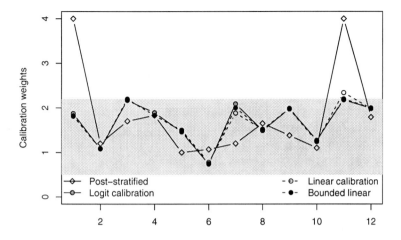

Figure 7.6   Impact of different calibration metrics. Artificial data from Kalton and Flores-Cervantes [69]. The gray rectangle shows the calibration bounds $(0.5, 2.2)$.

calibration methods. Code to reproduce their analyses is in the tests subdirectory of the survey package, in the file kalton.R and Figure 7.6 shows some of the resulting sets of calibration weights. The data form a 3×4 table and there are three sets of calibration weights using the margins of the table, and the set resulting from post-stratification on the cells of the table. The three sets of weights based on the margins of the table use linear calibration, logit calibration, and bounded linear calibration, with both logit and bounded linear calibration having a lower bound of 0.5 and an upper bound of 2.2, as indicated by the shaded region. The two bounded techniques give weights inside the shaded region; linear calibration gives one weight slightly outside the region at 2.34.

It is clear from the graph that the choice of auxiliary variables is much more important than the choice of calibration function. The three sets of weights using the margins of the table are very similar, and are quite different from the weights using all the cell counts in the table. Adding additional calibration variables will always give increased precision (or reduced non-response bias) with sufficiently large sample sizes, but for any given sample size there will be a point where the added uncertainty from estimating parameters in the calibration model outweighs the gains. Judkins et al. [68] describe one approach to assessing whether added auxiliary variables are helpful.

Cluster-level weights.   When the last stage of a probability design involves cluster sampling, such as sampling all individuals in a household, the sampling weights for individuals in the same cluster are necessarily identical. Especially for official statistics, it may be desirable for the calibrated weights for individuals in the same cluster to also be identical. This provides internal consistency between estimates, though at a small cost in efficiency. For example, in collecting data on

```
> cal2 <- calibrate(clus2_design, ~stype,
    pop=c(4421+755+1018,755,1018),
    aggregate.stage=1)
> svytotal(~enroll, cal2,na.rm=TRUE)
        total     SE
enroll 3084777 246321
> svytotal(~enroll, ps_design,na.rm=TRUE)
        total     SE
enroll 3074076 292584
> range(weights(cal)/weights(clus2_design))
[1] 1.075826 1.265473
> range(weights(cal2)/weights(clus2_design))
[1] 0.6418424 1.6764125
```

Figure 7.7   Forcing calibration weights to be constant within a school district

a sample of births, where every infant must have exactly one mother, it would be undesirable for estimated subpopulation totals of mothers to add up to more than the estimated total number of infants.

The aggregate.stage argument to calibrate() specifies the stage of sampling at which calibration weights must be constant within clusters. This argument is available only for survey design objects that include the sampling design. For survey design objects based on replicate weights the aggregate.index argument specifies a vector of cluster identifiers so that calibration weights will be made equal within the same cluster.

As an example, consider the two-stage cluster sample from Academic Performance Index population. The first stage of the design samples school districts and the second stage samples up to five schools within each district. Sampling probabilities, and thus sampling weights, are the same for schools within the same school district. When the sample was post-stratified on school type in Figure 7.1 the calibration weights were not the same for schools within the same cluster, varying up to 20% between school type. In Figure 7.7 the weights are forced to be constant within a school district, that is, within each stage-1 sampling unit. There is a slight cost in precision, with the standard error for the estimated total enrollment increasing by about 20%. This loss of precision occurs because the weights become more variable across school districts to compensate for the added constraints. The calibration weights in the simple post-stratified design range from 1.08 to 1.27, with constant weights within school district the range is from 0.64 to 1.68.

It is also possible to do the converse: to use auxiliary information where population totals are available only for sampled clusters at some stage of sampling, not for the whole population. This is done by supplying a list of population totals for the sampled clusters, with the names on the list matching the cluster identifiers, and giving the stage argument to calibrate() to indicate which stage of sampling the totals belong to. An example using the same two-stage sample from the Academic

Performance Index population is given in the `tests` subdirectory of the survey package, in the file `caleg.R`.

7.5 BASU'S ELEPHANTS

Basu [3, 97] gave an example that was intended to show unreasonable behavior of design-based inference, but that can be interpreted in terms of poor use of auxiliary information. In his story a circus owner had 50 elephants and wanted an estimate of their total weight based on weighing only one elephant. In the absence of any additional information a sensible approach would be to take a simple random sample of one elephant and weigh it (E_1), and then multiply by the sampling weight.

$$\hat{T}_{\text{srs}} = \frac{1}{\pi_1} E_1 = 50 \times E_1.$$

This is a valid design-based estimate, although no unbiased estimate of the standard error is available from a sample of size one.

The circus owner knows that five years previously, when all the elephants were last weighed, a particular mid-sized elephant, Sambo, had very nearly the average weight. A reasonable model-based estimate of the total weight of the elephants now would be

$$\hat{T}_{\text{model}} = 50 \times E_{\text{Sambo}}.$$

This is not a valid design-based estimate; no unbiased design-based estimate can be obtained when the sampling probabilities are zero for 49 of 50 elephants.

Basu imagines a compromise design worked out between the circus owner and the circus statistician, in which Sambo is sampled with very high probability ($\pi_{\text{Sambo}} = 0.99$) and the remaining probability is divided up among the other elephants ($\pi_i = 1/5000$). When the sampling is performed, Sambo is in fact chosen. The circus owner expects the estimate to be $50 \times E_{\text{Sambo}}$, but the statistician points out that the Horvitz–Thompson estimator is

$$\hat{T}_{\text{HT}} = \frac{1}{\pi_i} E_i = \frac{100}{99} \times E_{\text{Sambo}}.$$

This estimate is clearly silly. Worse still, if one of the other elephants, e.g., the largest elephant, Jumbo, had been weighed, the Horvitz–Thompson estimator would be

$$\hat{T}_{\text{HT}} = \frac{1}{\pi_i} E_i = 5000 \times E_{\text{Jumbo}}.$$

The Horvitz–Thompson estimator is exactly unbiased, but this is a property defined by averaging over possible realizations of the sampling design. For any single realization of the sampling design the result will be clearly unreasonable.

This example is somewhat embarrassing for design-based inference, so it is worth considering why such poor results are obtained. To some extent the unreasonable result is the fault of the small sample. In large samples we can be confident that the value of an estimate will be close to its expected value, so it is useful to know that the expected value is equal to the true population value. In small samples, where an estimate need not be close to the expected value, unbiasedness is not sufficient to ensure reasonable behaviour. However, the Horvitz–Thompson estimate is supposed to be useful even in small samples, so something is still wrong.

The first question is whether the "compromise" design is actually a sensible use of the prior knowledge if the Horvitz–Thompson estimator is going to be used. The second question is whether the Horvitz–Thompson estimator is appropriate, given the auxiliary information that is available. The answer to both questions is "No."

Using auxiliary information in design.   In choosing which elephant to weigh it is helpful to consider a situation where the sample size is slightly larger. If the circus were weighing three elephants rather than one, they could consider stratified sampling. Equation 2.6 in section 2.6 gives the optimal allocation: proportional to the number in each stratum and to the standard deviation in the stratum. If the elephants were divided into small, medium, and large by eye (or by previous weight), taking one elephant at random from each stratum would make sense. This leads to reasonable estimates and makes use of available information. A stratified sampling approach would lead to sampling probabilities π_i that did not vary much between elephants.

A similar approach, ranked-set sampling [72, 107, 123], is used in some ecological applications. Three samples of three elephants would be taken at random, and the elephants in each sample ranked by size. The smallest elephant from the first sample, the middle elephant from the second sample, and the largest elephant from the third sample are weighed. The ranking procedure has a similar effect to stratification, but ranking is often easier — lining up all the elephants by size would be more work than judging which of three is largest. Under ranked-set sampling the sampling weight π_i is the same for every unit in the population.

Section 3.3 considered sampling clusters proportional to size, or more generally, sampling observations proportional to some auxiliary variable available for the whole population. If the sampling probabilities are roughly proportional to the variable being analyzed, the variance of the estimated total will be small. In applying this principle to the elephants the circus owner could use either the weights recorded at the previous weighing, or if these are lost, an approximate measure of elephant volume as height × length × width. Under this PPS sampling scheme the sampling probability would be highest for the largest elephant. The variation between elephants in sampling probabilities would still be much smaller than for Basu's design, since the record weight for an elephant is only about twice the typical adult weight.

Based on the standard techniques used for designing complex samples it appears that if the Horvitz–Thompson estimator were to be used in analysis, a good design would have much less variation in sampling probabilities than Basu's design and that sampling larger elephants with modestly higher probability would be helpful.

Using auxiliary information in analysis.   A better way to use auxiliary information about the elephants would be by some form of calibration. For example, calibrating the estimated population size $N = 1/\pi_i$ to the known population size

$N = 50$ gives calibrated weights $g_i / \pi_i = 50$. The resulting model-assisted estimate of the total is

$$\hat{T}_{\text{cal:N}} = \frac{g_i}{\pi_i} \times E_i = 50 \times E_i$$

whichever elephant is sampled, the estimate that the circus owner wanted to use. This estimate was proposed by Hájek [57] in the discussion of Basu's original essay, based on the fact that dividing by the estimated (rather than known) population size often gives improved estimates of the population mean; it is also known as the Hájek ratio estimator of the total.

The estimate $\hat{T}_{\text{cal:N}}$ is not unbiased. It would be approximately unbiased for a large sample size, and in that sense is a design-based estimator, but 1 is not a large sample size. On the other hand, for almost any imaginable population configuration it will be better than $\hat{T}_{\text{HT}}$. Using this design and estimator will give more accurate estimates than taking a simple random sample and using $\hat{T}_{\text{srs}}$. Exercise 7.2 examines this issue by simulation.

It is possible to do better using the information from the previous weighing. If X_i is the weight of elephant i five years ago and T_X is the total of X, the weights can be calibrated on X giving the calibration constraints

$$\frac{g_i}{\pi_i} X_i = T_X.$$

Two solutions to the calibration constraints are

$$\frac{g_i^{(\text{ratio})}}{\pi_i} = \frac{T_X}{X_i}$$

and

$$\frac{g_i^{(\text{diff})}}{\pi_i} = \frac{T_X - 50 X_i}{X_i} + 50$$

corrresponding to the ratio working model

$$E_i = \alpha X_i + \epsilon_i$$

and the difference working model

$$E_i = \alpha + X_i + \epsilon_i.$$

The resulting estimates of the total are

$$\hat{T}_{\text{ratio:X}} = \frac{T_X}{X_i} E_i = \frac{E_i}{X_i} T_X,$$

where the previous total is multiplied by the relative increase in weight for the measured elephant, and

$$\hat{T}_{\text{diff:X}} = T_X + 50 \times (E_i - X_i)$$

where 50 times the absolute increase in weight for the measured elephant is added to the previous total.

These estimators are likely to be more precise than the Horvitz–Thompson estimator because they only need to estimate the relatively small change in weight since the previous weighing. Using these estimators it is no longer desirable to sample Sambo with higher probability, in contrast to $\hat{T}_{cal:N}$, because the auxiliary information is already being used efficiently.

The fact that only one elephant is weighed limits the choice of working models to the simple ratio and difference models. If the circus owner had wanted to weigh five elephants out of 250 it would be possible to do better using a working model with more parameters, such as

$$E_i = \alpha + \beta X_i + \epsilon_i.$$

This analysis shows that the circus statistician's failure in Basu's example was not adherence to design-based inference but ignorance of calibration estimators as the appropriate way to use auxiliary information. It is true that some of the approaches described here use more information than just the fact that

$$T \approx 50 \times E_{Sambo}.$$

If the circus owner knew that this was a sufficiently accurate approximation it would obviously be sensible just to weigh Sambo, but it is hard to imagine being sure of this without knowing anything else. For example, it is hard to imagine how the owner would know that the weight gains of the elephants over the past five years add up to 50 times Sambo's weight gain without also knowing that, e.g., the weight gains have been similar for each elephant.

The point of this example is not to argue that design-based estimates are to be preferred to model-based estimates, but to show that when a design-based estimate gives a clearly inappropriate estimate it is probably because it is not a good design-based estimate.

7.6 SELECTING AUXILIARY VARIABLES FOR NON-RESPONSE

Post-stratification, raking, and calibration are widely used to reduce the bias from *unit non-response*, people who cannot be contacted or refuse to participate in surveys. This problem is increasing over time, especially for telephone surveys; the University of Michigan's Survey of Consumer Attitudes found response rates decreasing by about 1% per year from 1979 to 2003 (Curtin et al. citecurtin-nonresponse), and the response rate in the Behavioral Risk Factor Surveillance System declined from about 70% in 1991 to about 50% in 2001. Reweighting for non-response is not a purely design-based method; it relies implicitly on models for the missing data.

In the simplest case, post-stratification, the model is that non-response is independent of the outcome variable within groups defined by the auxiliary variables. For example, suppose the probability of responding to a telephone survey about health insurance was higher for landline than cellphone users but within each group the probability of responding was independent of whether the individual had health insurance. An estimate using sampling weights based on the design would give too little weight to cellphone users (who tend to be younger) and so would be biased. If the telephone companies could be persuaded to give population numbers of landline and cellphone numbers, the analysis could be post-stratified by telephone type to give a valid estimate. This example also illustrates some of the limits of non-response adjustment: no reweighting, however sophisticated, will allow a telephone survey to give information about people without telephone service.

One way to look at the effect of post-stratification is that the correct sampling probability for a homogenous group of people is not the intended probability π_i but the achieved probability. If we try to sample 100 women aged 25–35 from a population of 10,000 and only 76 of them respond, the actual sampling fraction is not $100/10,000$ but $76/10,000$. This correction is exactly what post-stratification does

$$\frac{\pi_i}{g_i} = \pi_i \times \Pr[\text{response rate}].$$

There are two relevant ways that a group can be homogenous. If the non-response probability is homogenous within the group, as above, then post-stratification will give correct sampling weights, in the sense that the population total for any variable is correctly estimated. On the other hand, if a particular outcome variable is homogenous within the group, the population total for that outcome variable will be correctly estimated even if response is not homogenous: the relative weighting for individuals within the group will be wrong, but as the outcome is the same this incorrect relative weighting does not matter. More generally, if a regression model using the auxiliary variables can explain most of the variation in an outcome, the residual variation will be small and bias in analyzing it will be small in absolute terms even if large in relative terms. These two possibilities are analogous to the two constructions of calibration by "regression thinking" and "calibration thinking" in section 7.4.1.

For large-scale surveys there are usually only a small number of possible auxiliary variables and the goal must be to be produce correct results for all analyses, not just for a few chosen variables, so homogeneity of response is much more important. Of course, there is no realistic prospect of achieving true homogeneity of response using the few auxiliary variables typically available, but the estimates after post-stratification are probably less biased than those before post-stratification.

Keeter et al. [73] published results from an encouraging experiment that administered the same questionnaire in two telephone surveys, one of which made very extensive efforts to reduce non-response. The non-response rates were 36% and 60% for the two surveys. Even before any form of reweighting, the differences in political and social attitudes between responders to the two surveys wer much smaller than the differences in demographic variables. This shows that even quite high levels of non-response in an otherwise well-conducted survey may still give reasonable results. It also suggests that post-stratification or calibration should work well, since the demographic variables most likely to be used for reweighting the sample appeared more sensitive to response rates than the outcome variables being studied.

7.6.1 Direct standardization

Direct standardization of rates is the term used in epidemiology and demography for reweighting a sample from one population so that the distribution of variables such as age group and sex matches a different population. Direct standardization is used either to extrapolate to an estimate of the rate in the target population or to compare the extrapolated rate to the observed rate in the target population. For example, comparing the outcomes of neonatal care in different hospitals is difficult because the hospitals may have different numbers of high-risk, low birth weight infants. Direct standardization for birthweight allows a comparison based on the same distribution of birth weight in different hospitals.

Post-stratification for non-response and direct standardization are mathematically equivalent, but post-stratification is usually done with the intention of getting an improved estimate in the target population, and direct standardization is more often done to compare the known rates in the target population with rates extrapolated from the source sample. That is, post-stratification for non-response relies on the assumption that category-specific rates will be the same in the sample and target population, where direct standardization is often used to evaluate whether or not the category-specific rates are the same.

7.6.2 Standard error estimation

In theory, the same approaches to standard error estimation that apply to post-stratification and calibration for precision also apply to post-stratification and calibration for non-response. When conducting secondary analysis of large-scale surveys it is often not possible to use the correct standard error estimates, because the information needed to compute the residuals is not published. In these cases, rather than computing the standard errors from residuals as in equation 7.1 and 7.4, the standard errors are computed as if the calibrated weights g_i / π_i are simply the sampling weights. This approximation, like the single-stage approximation for multistage sampling, is typically conservative.

When survey data are published with replicate weights, as in the California Health Interview Survey, it is possible to produce the correct standard errors by ensuring that each set of replicate weights is post-stratified or calibrated appropriately. Alternatively, if the non-response adjustments to the weights are sufficiently straightforward it may be possible to reproduce them at the time of analysis, as in the example of the Family Resources Survey in section 7.3.

EXERCISES

7.1 Using the Washington State Crime population, take a stratified random sample of five police districts from King County and five counties from the rest of the state.

   a) Calibrate the sample using stratum and population as the auxiliary variables. Estimate the number of murders and number of burglaries in the state using the calibrated and uncalibrated sample.

b) Convert the original survey design object to use jackknife replicate weights (with as.svrepdesign()). Calibrate the replicate-weight design using the same auxiliary variables and estimate the number of burglaries and of murders in the state.

c) Calibrate the sample using the population and the number of burglaries in the previous year as auxiliary variables, and estimate the number of burglaries and murders in the state.

d) Estimate the ratio of violent crimes to property crimes in the state, using the uncalibrated sample and the sample calibrated on population and number of burglaries.

7.2 Write an R function that accepts a set of 50 elephant weights and simulates repeatedly choosing a single elephant and computing the Horvitz–Thompson and ratio estimators of the total weight, reporting the mean and variance over the repeated simulations. Explore the behavior for several sets of elephants weights. Verify that the Horvitz–Thompson estimator is always unbiased, but that it is usually further from the truth than the ratio estimator.

7.3 ★ Write an R function that accepts a set of 50 elephant weights and performs the ranked-set sampling procedure on page 150 to choose three of them. By simulation, compare the bias and variance of the estimated total from ranked-set sample to estimated totals from a simple random sample of three elephants.

7.4 Estimate the proportions of people in California with normal weight, over-weight, and obesity using the BRFSS 2007 data (California is X_STATE = 6, and BMI categories are X_BMI4CAT). Post-stratify the California data to have the same age and sex distribution (X_AGE_G and X_SEXG_) as the data for Florida (X_STATE = 12) and compute the directly standardized estimates of proportions for the BMI categories. Compare the raw and standardized estimates based on California data to estimates from the data for Florida to see if the differences in BMI between the states are explained by differences in age distribution. [You will need to load the data subset for California directly into memory, as postStratify() does not currently support database-backed designs.]

7.5 ★ Consider a categorical post-stratification variable with K categories having population counts $N_1, N_2, \ldots, N_k$. Suppose we are interested in estimating the total of a variable Y

a) Show that the post-stratified estimate is

$$\hat{T}_{ps} = \sum_{k=1}^{K} N_k \hat{\mu}_k$$

where $\hat{\mu}_k$ is the estimated mean of Y in group k before post-stratification.

b) Show that the regression estimate from a model with indicator variables for each group is also

$$\hat{T}_{\text{reg}} = \sum_{k=1}^{K} N_k \hat{\mu}_k.$$

CHAPTER 8

TWO-PHASE SAMPLING

In which we take one step at a time.

8.1 MULTISTAGE AND MULTIPHASE SAMPLING

Multistage sampling, as described in Chapter 3, requires the assumption that the subsampling of one PSU does not depend on which other PSUs were sampled. It is possible to relax this assumption and to allow subsampling that depends on all the currently observed data at each step, a process called *multiphase sampling*.

This chapter discusses two scenarios where two-phase sampling is helpful in designing subsamples. There are many other possible two-phase and multiphase sampling designs, including longitudinal designs where overlapping samples are taken on multiple occasions, and dual-frame designs where samples are taken from two incomplete population lists. At the time of writing these other designs are not directly supported in the survey package, although it should be possible to translate instructions for analyzing them using other software into R.

The distinction between two-stage and two-phase subsampling is important theoretically, although it has relatively little impact on applied analysis of subsampling designs. The reason for the theoretical distinction is that it is not possible to calculate

Complex Surveys: A Guide to Analysis Using R. By Thomas Lumley
Copyright © 2010 John Wiley & Sons, Inc.

the true sampling probability for an individual. We have available the probability $\pi_{i(1)}$ that individual i is in the first-phase sample, and the probability $\pi_{i(2|1)}$ that that the individual is then included at phase two. For multistage sampling, in Chapter 3 we could simply multiply the stage-one and stage-two probabilities to obtain π_i, because of the assumption that subsampling for one cluster did not depend on anything outside that cluster. The purpose of two-phase designs is to relax this assumption so the subsampling probability can be based on the composition of the first-phase sample and could be different for other first-phase samples. In order to compute the probability that individual i is in the sample we would need to average $\pi_{i(1)} \times \pi_{i(2|1)}$ over all possible first-phase samples that include individual i, which typically requires data we do not have as well as being computationally infeasible.

Fortunately, the numbers $\pi_i^* = \pi_{i(1)} \times \pi_{i(2|1)}$ can be used in the same way as π_i to create sampling weights that give unbiased estimators of the population total, and the corresponding pairwise numbers can be used to estimate standard errors. The resulting estimator is not the Horvitz–Thompson estimator, but it looks, walks, and quacks like it. The mathematical details of estimation are given in Chapter 9 of Särndal et al. [151], much of which follows Särndal and Swensson [150].

Two-phase samples can be described to R with the function `twophase()`. At the time of writing the designs that can be handled are relatively restricted. Either phase one is a stratified sample of individuals, or phase one is a cluster sample in which all clusters are represented at phase two. The call to `twophase()` has similar arguments to the call to `svydesign()`, but each argument is a list of two formulas rather than a single formula, representing the two phases of sampling. More detail is given in the examples in section 8.4

8.2 SAMPLING FOR STRATIFICATION

Two-phase sampling for stratification can be used when a valuable stratification variable is not available for all individuals in the population but can be measured inexpensively. The strategy is to take a large sample from the population, measure the stratification variable, and then take a stratified subsample. The phase-one sample can be a simple random sample or can be stratified on other variables that are available for the population. If the phase-one sample is sufficiently large, the distribution of the stratifying variable across the phase-one sample will be very similar to the distribution across the population, and the design will give very similar estimates to the stratified one-phase design it is emulating (Exercise 8.1).

One situation where two-phase sampling for stratification has been popular is in estimating the age distribution of fish caught by the commercial fishing industry. Kutkuhn [88] describes a two-phase sampling system for the California salmon catch, where fish size is measured on a large sample of fish at the port and then used to choose a stratified subsample of fish to have age determined accurately by counting rings on scales. Smith [164] describes a similar system for estimating the age distribution of halibut, and concludes that the phase-one sampling is sufficiently expensive that a simple random sample would be more cost-effective.

Screening of a phase-one sample for stratification in psychiatric research is described by Pickles et al. [126]. A simple questionnaire is administered at phase one, and the results are used to draw a stratified subsample with high sampling probabilities where the screening questionnaire indicated a high chance of mental illness.

NHANES and NHIS use a screening design based on ethnicity in the third and fourth stage of their multistage sample (Mohadjer and Curtin[109]). After small clusters of households have been chosen in a two-stage design, some household clusters are screened and included in the sample only if they include one or more black, Asian, or Hispanic persons. This two-phase sampling is not represented in the public-use data files and so cannot be incorporated in any secondary analysis. I do not know if the internal analyses at the National Center for Health Statistics explicitly incorporate the two-phase structure of the sample.

8.3   THE CASE–CONTROL DESIGN

Probably the most important example of sampling for stratification is the classical *case–control design* (or *choice-based design* in economics). The case–control design is used in epidemiology to study rare diseases such as cancers, where any of the sampling designs we have seen so far would end up sampling very few people with the disease.

The case–control design assumes that it is relatively easy to identify all the cases of the disease in some group of people and, implicitly, to identify all the controls, ie, people without the disease. The first phase of sampling results in the group of people whose disease status is known. In the second phase all the cases are sampled, but only a small fraction of the controls — typically 1–5 times as many controls as cases. Predictor variables are then measured on the second-phase sample. If the cases make up 1/1000 of the population and the same number of cases as controls is used, the sampling fractions will be 1 for cases and 1/1000 for controls. The design effect for a case–control sample is of the same order of magnitude as the sampling probability for controls; the design is enormously more efficient than simple random sampling.

When the cases come from a well-defined population and the controls genuinely are a random sample of the same population, the design is called *population-based*. For example, the cases may be all the stroke cases at a particular HMO and the controls a sample from HMO members, or the cases may be all the cancer cases in a particular state and the controls a sample from the population of that state. It is often hard to establish a population for the cases, and so controls are often sampled in ways that only approximate the formal case–control design — not surprisingly, problems with the choice of controls are a common criticism of these designs. A historical review of the case–control design is given by Breslow [12].

In fact it is unusual for design-based inference to be used in the case–control design. The usual analysis relies on the fact that one particular association measure, the odds ratio, can be estimated correctly without using sampling weights as long as the logistic regression model is correctly specified. With a single binary predictor variable this result is elementary (Table 8.1): the odds of exposure in cases does

Table 8.1   The odds ratio is $(a/b)/(c/d) = ad/bc$ under simple random sampling and $(a/b)/((c/\pi_0)/(d/\pi_0)) = ad/bc$ under case–control sampling with π_0 sampling fraction in controls.

	Exposed	Unexposed			Exposed	Unexposed
Case	a	b		Case	a	b
Control	c	d		Control	c/π_0	d/π_0
Total	$a+c$	$b+d$		Total	$a+c/\pi_0$	$b+d/\pi_0$

not depend on the controls in any way, and the odds of exposure in controls does not depend on the sampling fraction for controls because all controls have the same sampling probability. The general result for discrete variables was given by Anderson [1] and for continuous variables by Prentice and Pyke [128].

In addition to simplicity of estimation and the fact that it can be applied without knowing the population sizes, the unweighted estimator always has the same or smaller standard errors. In some situations the gain in precision can be quite large, for example, Scott and Wild [158] showed simulations where the weighted estimator would need up to eight times the sample size to obtain the same precision. On the other hand, the difference in precision is small when the odds ratio is not far from 1, and simulations using different distributions from those used by Scott and Wild show smaller differences in precision even at large odds ratios. Some of these simulations are presented in section 8.3.1.

The disadvantage of the unweighted estimator is the fact that its validity relies on the logistic regression model being correct, something that many statisticians are reluctant to assume. As Scott and Wild[158] observe

> The usual practice when building parametric regression models is to start with simple candidate surfaces and to complicate them only when the data force us to do so. As a consequence, we are always working with a model that is not quite right.

We saw in Chapter 5 that the weighted estimator always estimates a well-defined quantity, the best-fitting population logistic model, even if the model is not exactly correct. The potential for being misled by bias is less serious than in the scenarios in Chapter 5. Using the weights gives the best-fitting logistic regression model in the population, omitting the weights gives the best-fitting regression model in a modified population where the risk of disease is higher at any covariate value [158].

In the past this issue has been quite controversial in statistics. It now seems that both the bias from a misspecified model and the precision gain from avoiding weights are smaller than they were feared to be. For the relative small effect sizes that are most common it appears that there is no compelling argument in either direction and a reasonable approach might be to fit both the weighted and unweighted estimators.

Example: Oesophageal cancer in Ille-et-Vilaine.   Breslow and Day[14] used as an example a case–control study of oesophageal cancer conducted in the Ille-et-Vilaine region of northwest France by Tuyns et al. [176].

The control sampling probability for these data was not reported by Tuyns et al. [176], but in an earlier paper [175] reporting cancer incidence rates they gave the population of the *département* of Ille-et-Vilaine as 430,000, implying a control sampling probability of 0.00225, or a sampling weight of 441. As the phase one sample is large, treating the design as a single phase of stratified sampling with replacement gives an essentially identical answer to using the full two-phase representation, so it is not necessary to construct a data set with the 429,000 controls in Ille-et-Vilaine who were implicitly in the phase-one sample. Exercise 8.2 does construct the full two-phase representation to confirm that the same result is obtained.

The data are built in to R, in a data set called esoph, but in a tabular form listing the number of cases and controls for 88 combinations of predictors. The code in Figure 8.1 expands the data to one record for each person and creates the weight variable, then performs weighted and unweighted logistic regressions. The model, which fits the data well, has linear terms in the grouped tobacco and alcohol variables and a separate indicator variable for each age group.

The estimated log odds ratios are very similar in the two models, and highly statistically significant. The standard errors from the unweighted model are 5% smaller for the tobacco coefficient and essentially identical for the alcohol coefficient. In this example there is little inefficiency from using the weights, but also little difference in the estimates, and the same inference would be drawn from either analysis.

8.3.1 ★ Simulations: efficiency of the design-based estimator

There are two approaches to simulating a case–control design. The simplest is to simulate the phase-one sample and subsample from it. This tends to be slow, especially on computers with limited memory, when the sampling fraction for controls is small. For example, simulating the Ille-et-Vilaine case–control design would require repeatedly simulating a population of size 430,000. The other approach is to specify the covariate distribution in controls and then use the relationship

$$\Pr[X = x | Y = 1] \propto e^{x\beta} \times \Pr[X = x | Y = 0].$$

For discrete covariates this *exponential tilting* procedure gives the case distribution directly. For continuous covariates from an exponential family distribution, in particular the Normal distribution, it is also straightforward to compute the distribution in cases. The second approach is less general, since it can be used only when the true relationship is logistic and when the predictor distribution in controls makes exponential tilting easy, and it is difficult to adapt to complex sampling at phase one.

When examining the efficiency of the weighted estimator the extra generality provided by simulating the entire first-phase sample is not necessary. The code in Figure 8.2 creates Normal or categorical data and fits the logistic regression model using both the weighted and unweighted estimators. It then performs 1000 replications of a simulation with $\beta = 0.5$, 500 cases and 500 controls, and a control sampling fraction of 1/1000, for three data distributions. The first is a Normal distribution, the second is a square distribution taking values 1–4 with equal probability in controls,

```
> cases <- cbind(esoph[rep(1:88, esoph$ncases), ],
      case=1, weight=1)
> controls <- cbind(esoph[rep(1:88, esoph$ncontrols), ],
      case=0, weight=441)
> esoph.x <- rbind(cases,controls)
> d_esoph <- svydesign(id=~1,  strata=~case, weights=~weight,
      data=esoph.x)
> unwtd <- glm(case~agegp+as.numeric(tobgp)+as.numeric(alcgp),
        data=esoph.x, family=binomial )
> wtd <- svyglm(case~agegp+as.numeric(tobgp)+as.numeric(alcgp),
        design=d_esoph, family=quasibinomial)
> coef(unwtd)[7:8]
as.numeric(tobgp) as.numeric(alcgp)
        0.2616223                0.6530835
> coef(wtd)[7:8]
as.numeric(tobgp) as.numeric(alcgp)
        0.2342193                0.6067380
> SE(unwtd)[7:8]
as.numeric(tobgp) as.numeric(alcgp)
        0.08197699              0.08452042
> SE(wtd)[7:8]
as.numeric(tobgp) as.numeric(alcgp)
        0.08588926              0.08374864
```

Figure 8.1   Case–control study of oesophageal cancer in Ille-et-Vilaine

```
make.normal.data<-function(n, beta,pi0){
   xctrl<-rnorm(n)
   xcase<-rnorm(n, m=beta)
   data.frame(x=c(xctrl,xcase),y=rep(0:1,each=n),
      w=rep(c(1/pi0,1),each=n))
}

make.categorical<-function(n, beta, pi0, probs){
   k<-length(probs)
   xctrl<-sample(1:k,n, replace=TRUE, prob=probs)
   caseprobs<- probs*exp(beta*(1:k))
   caseprobs<-caseprobs/sum(caseprobs)
   xcase<-sample(1:k,n,replace=TRUE, prob=caseprobs)
   data.frame(x=c(xctrl,xcase),y=rep(0:1,each=n),
      w=rep(c(1/pi0,1),each=n))
}

estimate<-function(data){
    des <- svydesign(id=~1,strata=~y, weights=~w, data=data)
    m1 <- glm(y~x, data=data, family=binomial())
    m2 <- svyglm(y~x, design=des, family=quasibinomial())
    c(coef(m1), coef(m2), SE(m1), SE(m2))
}

n0.5<-replicate(1000,
   estimate(make.normal.data(500, 0.5, 1/1000)))
squareprob<- c(1/4,1/4,1/4,1/4)
sq0.5<-replicate(1000,
   estimate(make.categorical(500, 0.5, 1/1000, squareprob)))
triprob<- (4:1)/10
tr0.5<-replicate(1000,
   estimate(make.categorical(500, 0.5, 1/1000, triprob)))

apply(n0.5,1,mean)
apply(n0.5[1:2,],1, sd)
```

Figure 8.2   Case–control simulation code

and the third is a triangular distribution taking values 1–4 with probabilities 0.4, 0.3, 0.2, 0.1 in controls. The calls to apply() compute the means and standard errors of the estimators.

Table 8.2 shows the results of running these three simulations and repeating them for $\beta = 0, 0.5, 1, 1.5, 2$. The number in each cell of the table is the relative efficiency of the weighted estimator. If the relative efficiency is 50%, the weighted estimator

Table 8.2   Relative efficiency of weighted estimator in case–control design

Distribution	$\beta = 0$	$\beta = 0.5$	$\beta = 1$	$\beta = 1.5$	$\beta = 2$
Normal	100	85	48	20	22
Square	100	96	94	92	97
Triangular	100	94	84	76	77

requires twice as many observations for the same accuracy. The results for the Normal distribution agree with those of Scott and Wild [158], but the results for the two discrete distributions show much less difference between the two estimators.

The distribution of alcohol and tobacco intake for the Ille-et-Vilaine oesophageal cancer study is similar to the triangular distribution in the simulations. The efficiency of the weighted estimator is slightly higher for the Ille-et-Villaine data set than in the simulations because there are approximately five times as many controls as cases.

8.3.2 Frequency matching

Many case–control designs use a further level of stratification and unequal sampling in the phase-two sample, a practice known in the epidemiology literature as *frequency matching*. The goal of frequency matching is to prevent small case vs control differences in the exposure under study from being masked by very large differences in another variable whose effects are known and not of interest.

For example, in the oesphageal cancer study described above, very little information about the effects of alcohol and tobacco is present in the youngest age group, because there is only one case. If the associations with age had already been understood and the study had been designed to estimate the effect of alcohol and tobacco this would be an inefficient design that effectively wasted 116 controls. A more efficient design would sample more controls at older ages and end up with five controls per case in each age group rather than five controls per case on average over all ages.

Frequency matching is not the most efficient way to use the age information in a two-phase design, but it has the advantage that an unweighted analysis is still valid as long as the stratification variables are included in the model. The coefficients for the stratification variables will not be correct, but these are assumed to be uninteresting. The coefficients for other variables will be correct under the same assumption of a correct model that was made for a unmatched case–control design.

8.4 SAMPLING FROM EXISTING COHORTS

In two-phase sampling for stratification there is very little information available at phase one, and most variables are measured only at phase two. The opposite situation is rare in official statistics but occurs in epidemiologic studies when new variables are measured on a subset of an existing sample; many variables at phase one and only a little extra information at phase two.

Large observational cohort studies and randomized trials recruit a sample of thousands of people (treated for analysis purposes as a simple random sample from a very large population) and measure hundreds of variables over years or decades of follow-up. As new research questions and measurement techniques arise, investigators need to measure new variables. These may be new assays of stored blood and DNA samples, data from reviews of medical records, or simply coding and entry of free-text questions on, eg, nutritional supplements or over-the-counter medications.

Measuring these new variables is often expensive, so restricting the new measurements to a subsample of the full cohort is desirable. Rather than taking a simple random sample, it is more efficient to use the existing variables to stratify the sampling. The traditional approach was to stratify on a single outcome variable, giving a nested case–control or case–cohort [127] design. Data from phase one for individuals not in the phase-two sample was ignored.

More recently, epidemiologists have become aware of the potential advantages of stratifying the phase-two sample on predictors as well as response, and of poststratifying or calibrating the phase-two sample to the full cohort. As with the nested case–control design there are sampling-weighted estimators that do not rely on the assumption that a model has been correctly specified, and semiparametric maximum likelihood estimators that do make this assumption. The relative importance of the bias from assuming the model is correct and the efficiency loss from not assuming it is correct are even less well understood than in the nested case–control design, and two-phase model-based estimation is an area of active research at the time of writing.

8.4.1   Logistic regression

In the 2×2 table in Table 8.1 the estimated variance of the log odds ratio is

$$\operatorname{var}[\log \psi] = \frac{1}{a} + \frac{1}{b} + \frac{1}{c} + \frac{1}{d}.$$

The case–control design ensures that the number of cases, $a + b$, and the number of controls, $c + d$, are both large, but if exposure is rare it is still possible for a or c to be small. For example, consider the data in Table 8.3, from the National Wilms Tumor Study Group [54, 38]. All the participants in these studies have Wilms' tumor, so the 'cases' and 'controls' are cases of relapse and controls who did not experience relapse. A case–control sample of all 669 cases and 669 of the controls from this population would, on average, include only 50 controls with unfavorable histology.

If sampling could be based on both exposure and outcome it would be possible to have 100% sampling of the three smaller cells, plus 424 controls with good histology, adding up to the same sample size of $669 \times 2 = 1338$. Increasing the smallest cell count from 50 to 194 will clearly give better estimates of association. Of course, sampling stratified on the cells in Table 8.3 would require knowing histologic classification and relapse status for everyone in the population and sampling would

Table 8.3 Histologic classification and relapse in Wilms' tumor

| | Histology | | |
	Unfavorable	Favorable	Total
Relapse	194	475	669
Control	245	3001	3246
Total	439	3476	3915

then be unnecessary. The question this example raises is whether the potential gains from sampling based on exposure and outcome are ever available in practice.

There are at least two situations where the increased precision can be realized in practical designs. The first is when exposure and outcome are available for all of a phase-one sample and subsampling is being used to measure additional variables that may confound or interact with the exposure. The second is when the true exposure is not available at phase one but there are variables available that predict the outcome reasonably well.

The Wilms' tumor population data have been widely used as an artificial example of sampling based on a surrogate for exposure; the accounts here are based on Breslow and Chatterjee [13] and Kulich and Lin [85]. The histologic classification in Table 8.3 is performed at the central laboratory of the National Wilms Tumor Study Group, by the pathologist who first characterized the histologic variations of Wilms' tumor. There is also available a classification by pathologists at the hospital where treatment was performed. These pathologists will be less familiar with Wilms' tumor, which is a very rare disease, and their ratings might well be less accurate. Analyzing the complete data confirms this possibility. The central lab classification predicts relapse rate more strongly than the local hospital classification, and if the central lab classification is known there is little or no improvement in prediction of relapse by using the local hospital classification in addition. That is, essentially all the differences between the two sets of histologic classifications are errors by the local hospital pathologists, who have about 98% specificity but only about 75% sensitivity in detecting the unfavorable histology.

If it became desirable to restrict the central lab evaluation to retrospective classi-fication of a subsample of tumors, an obvious set of strata would be the four cells formed by tabulating relapse and local hospital histology. Sampling everyone with unfavorable (local hospital) histology, all the relapse cases with favorable histology, and 449 controls with favorable (local hospital) histology, gives the same sample size of 1138 as a 1:1 case–control sample. Table 8.4 shows the expected table of (central lab) histology by relapse. The smallest cell is controls with unfavorable histology, as for the case–control sample, but the expected number is 183 rather than 50.

The price for the increased flexibility in sampling is that ordinary logistic regres-sion no longer gives a valid analysis: the unweighted odds ratio in Table 8.4 is 1.08 and the true population odds ratio from Table 8.3 is 5.0. It is straightforward to esti-mate the odds ratio using weights based on π_i^*: the subsampling probabilities $\pi_{i(2|1)}$

Table 8.4 Central lab histologic classification and relapse in a subsample of Wilms' tumor data chosen based on relapse and local hospital histologic classification

	Histology Unfavorable	Favorable	Total
Relapse	194	475	669
Control	183	486	669
Total	377	961	1338

are known and since the complete sample is being modeled as a simple random sample from an infinite superpopulation all $\pi_{i(1)}$ are equal and can be taken as $\pi_{i(1)} = 1$. The weighted analysis also has the advantage of being able to estimate statistics other than odds ratios, such as the relative risk of relapse with poor histology or the sensitivity and specificity of the local hospital histology classification. Exercise 8.4 shows by simulation that the true sampling probabilities π_i, which depend on the superpopulation model, are likely to be very close to π_i^* for this design.

8.4.2 Two-phase case–control designs in R

A two-phase design is specified with the twophase() function. The syntax is similar to that for svydesign(), except that two sets of information are required, one for each phase. The id and strata arguments are lists of two model formulas, the first for phase one, the second for phase two. In Figure 8.3 the id argument specifies sampling of individuals at each phase. The strata argument specifies that the first stage is unstratified and the second stage is stratified on the cross-classification of instit and relaps. The subset argument is a logical (TRUE/FALSE) vector indicating which individuals are in phase two. The data are supplied as a data frame with a record for every individual, whether or not they are in phase two. If they are not in the phase two sample they would usually have NA values for any phase-two variables. A single set of sampling weights is supplied, which is $1/\pi_i^*$ for the individuals in the phase-two sample. The values for individuals not in the phase-two sample are not used, and could be set to NA if they are not known. When the design object is printed, the output shows the two sampling designs. Note that the phase-two design in Figure 8.3 has a fpc argument, even though none was supplied in the call to twophase(). The "population" for the phase-two sample is just the phase-one sample, so the population size in each stratum is known from the supplied data.

Figure 8.4 gives code for fitting a logistic regression model to the two-phase sample and to a case–control sample using both model-based and design-based analyses, and fitting a relative risk regression to the two-phase and nested case–control samples. The additional variables are stage, the stage of the tumor, or how much it has spread (0–3); tumdiam, the size of the tumor, and age at diagnosis. The estimates and standard errors are shown in Table 8.5.

```
> nwts <- read.table("nwts-share.txt", header=TRUE)
> set.seed(1337) # to make this reproducible
> subsample <- with(nwts, c(which(relaps==1 | instit==1),
        sample(which(relaps==0 & instit==0), 449)))
> nwts$in.subsample <- (1:nrow(nwts)) %in% subsample

> nwts_design <- twophase(id=list(~1, ~1), subset=~in.subsample,
    strata=list(NULL, ~interaction(instit, relaps)), data=nwts)
> nwts_design
Two-phase design: twophase(id = list(~1, ~1), subset =
 ~in.subsample, strata = list(NULL,~interaction(instit, relaps)),
 data = nwts)
Phase 1:
Independent Sampling design (with replacement)
svydesign(id = ~1)
Phase 2:
Stratified Independent Sampling design
svydesign(id = ~1, strata = ~interaction(instit, relaps),
fpc = '*phase1*')
```

Figure 8.3   Declaring a two-phase design for a sample of the NWTS population

```
nwts <- read.table("nwts-share.txt", header=TRUE)
set.seed(1337)
casectrl <- with(nwts, c(which(relaps==1),
        sample(which(relaps==0), 669)))
nwts$in.ccs <- (1:nrow(nwts)) %in% casectrl
ccs_design <- twophase(id=list(~1, ~1), subset=~in.ccs,
    strata=list(NULL, ~relaps), data=nwts)

m1 <- svyglm(relaps~histol*stage+age+tumdiam,design=nwts_design,
    family=quasibinomial())
m2 <- svyglm(relaps~histol*stage+age+tumdiam,design=ccs_design,
    family=quasibinomial())
m3 <- glm(relaps~histol*stage+age+tumdiam,data=nwts,
    subset=in.ccs, family=binomial())
m1a <- svyglm(relaps~histol*stage+age+tumdiam,design=nwts_design,
    family=quasipoisson(log))
m2a <- svyglm(relaps~histol*stage+age+tumdiam,design=ccs_design,
    family=quasipoisson(log))
```

Figure 8.4   Fitting logistic regression models to a sample of the NWTS population

Table 8.5   Results from NWTS analyses. Design A is sampled on `instit` and `relaps`, design B is the case–control design sampled on `relaps`.

Design	log(odds ratio) Design-based A	B	Model-based B	log(relative risk) Design-based A	B
$\hat{\beta}$					
(Intercept)	-2.664	-2.625	-1.061	-2.679	-2.63
histology	0.412	-0.02	0.154	0.676	0.405
stage	0.259	0.18	0.179	0.225	0.163
age	0.044	0.1	0.097	0.032	0.072
tumor diameter	0.008	0.001	0.003	0.006	0.001
histologyl:stage	0.481	0.652	0.592	0.163	0.267
Standard error					
(Intercept)	0.214	0.201	0.204	0.170	0.160
histology	0.382	0.474	0.438	0.25	0.290
stage	0.067	0.065	0.061	0.055	0.053
age	0.024	0.024	0.023	0.0170	0.017
tumor diameter	0.017	0.016	0.016	0.013	0.012
histology:stage	0.158	0.197	0.184	0.089	0.100

The first thing to note about the result is that the relative risks and odds ratios are quite different, especially for the histology×stage interaction. The reason is that relapse is quite common at advanced disease stages or with unfavorable histology. The intercept estimate also differs substantially between the design-based and model-based logistic regression analyses: the model-based estimate by ordinary logistic regression does not give the correct intercept. The bias should be equal to the log odds of being sampled for controls, which gives $\log(3246/669) = 1.58$. Correcting this bias gives an intercept of -2.84, in good agreement with the design-based analyses.

For both the log odds ratio and the log relative risk there is a noticeable gain in precision for interaction between histology and stage when sampling uses the surrogate exposure variable `instit`. There is relatively little impact on the standard errors of the other coefficients. The gain in precision is larger than the gain from using a model-based analysis of the case–control sample, and comes without having to make any additional assumptions.

As mentioned above, model-based approaches are also available for fitting linear and generalized linear models to two-phase samples. As with the case–control design, there is the potential for more precise estimates from the model-based approach and also the potential for bias when the model is misspecified. There has been a lot of theoretical research in this area, relatively little of which is published in ways accessible to the non-specialist. The most general implementation of model-based analysis for two-phase studies is that of Scott and Wild [159], with R software and examples available from http://www.stat.auckland.ac.nz/~wild/software.html.

8.4.3 Survival analysis

The case–cohort design was developed by Prentice [127] as an alternative to a nested case–control design in a cohort study. In the original case–cohort design the phase-two sample consists of a subcohort chosen at the beginning of the study, augmented over time by all those who became cases. The second phase is thus a stratified sample, stratified by case status. In contrast to the case–control design, where a given set of controls cannot readily be used for a different set of events, the same subcohort in a case–cohort can be used for studying cases of different diseases. Stratification on exposure variables is a natural extension, and was explored by Borgan *et al*[11] and by Samuelsen al. [146]. Survival analysis based on case–cohort designs is currently more widely used than logistic regression based on two-phase samples. Recently published examples include examples of genotyping of stored DNA[90, 91], new biochemical assays of stored blood [55, 70, 114], and new interviews and sample collection on the phase-two sample [87, 165].

The case–cohort design is intended for studying time to an event such as relapse or death. In a regression model where the outcome variable is time to an event there is the complication that this event will usually not be observed for everyone in the cohort. If someone is alive at the end of follow-up their time to death is not known, but a lower bound is known, the current duration of follow-up. In this situation we say that the time to death is *censored* at the end of follow-up.

It is not possible to estimate the probability of the event or the mean time until the event from censored data. Analyses of events and time-to-event in a cohort are most often based on the Cox proportional hazards model. This is related to the logistic regression model, but models the *rate* of occurrence of events over time rather than the risk or probability of occurence. The rate (or "hazard") for person i is written $h_i(t)$, and by historical convention predictor variables are generically called z rather than x. The working model specifies the rate as a function of time

$$\log \text{rate} = \log h_i(t) = \log h_0(t) + \beta \times z \tag{8.1}$$

where $\log h_0(t)$ is a set of intercepts for each time point, or equivalently

$$\text{rate} = h_i(t) = h_0(t) \times e^{\beta z}$$

In this working model the effect of z on rate is multiplicative and does not change with time, leading to the name 'proportional hazards model'. The numbers e^{β} are called *hazard ratios*, so the coefficients in the model, $\hat{\beta}$, are the estimated log hazard ratios. The collection of intercepts, $h_0(t)$ is called the *baseline hazard*.

The model in equation 8.1 does not explicitly mention the outcome variable. Operationally, the time to event outcome is represented by a combination of two variables in the data set: the duration of follow-up and the status at the end of follow-up. In R these variables are packaged together with the Surv() function when they are supplied to a model formula. Surv(time, status) represents observations with a follow-up duration of time and with status coded as 1 for those who had the event at the specified time and 0 for those who still had not had the event. For detailed introductions to survival analysis see Breslow and Day [15] or Kleinbaum and Klein [76].

The original methods for fitting the Cox model to case–cohort samples were designed so that the mathematical techniques available at the time could be used to study them. These mathematical techniques (Prentice[127], Self and Prentice [160])required the weight for each observation at each point in time to depend only on information collected before that point in time, even though data from the entire follow-up period was available. Therneau & Li [168] describe how to implement these analyses in R and other statistical software. Analyses based on a Cox model with two-phase sampling weights have now replaced the original methods. Design-based methods for fitting the Cox model to complex samples were first described by Binder [9]and with more mathematical detail by Lin [95]. The asymptotic theory is substantially more complicated than for generalized linear models, and is still an active area of research, especially for two-phase designs [20, 95].

8.4.4 Case–cohort designs in R

The Wilms' tumor population was used above in an example of logistic regression, but the time to relapse is also available, allowing case–cohort analyses to be simulated. For the classical case–cohort analysis we take a sample from the cohort at the start of followup and then add all the cases to it. If the expected event rate is about 1 in 7, giving about 650 expected cases, and we want about 650 non-cases, this means sampling a subcohort of $650 \times 7/6$ or about 750. For the one realization of the sampling scheme in Figure 8.5 there are 131 cases and 619 non-cases in the subcohort. Under this sampling design there are two sampled strata: those in the subcohort and those not in the subcohort. The sampling probabilities are $\pi_{(2|1)} = 750/3915$ for the subcohort and $\pi_{(2|1)} = 1$ for cases not in the subcohort.

In Figure 8.5 the svycoxph() function fits the Cox proportional hazards model after the two-phase design has been set up. The code also shows the classical model-based analysis of Prentice [127] for these data. The cch() function (in the survival package) differs from svycoxph() and twophase() in that its input data are just the observations in the second-phase sample. The total cohort size is supplied as an additional argument. The two analyses do not give the same results even for point estimates. For example, the coefficient for histol is 0.96 with standard error 0.38 from cch() and 0.75 (std error 0.35) from svycoxph(). The value using the entire sample is 0.59. The differences between methods are unusually large in this example because the Cox model does not fit particularly well. Figure 8.6 shows an estimate of the coefficient of age as a function of time, produced with plot.cox.zph(). The association is much stronger in the short term than later in time; the hazards are not proportional. The same is true for stage.

The different weighting for cases inside and outside the subcohort is aesthetically unattractive and will also result in a slight loss of efficiency. We could instead have sampled all the cases and taken a sample of 619 non-cases, giving a two-phase design stratified on case status. Equivalently, we could post-stratify on case status. Either approach would result in weights of 3246/619 for non-cases and 1 for cases.

```
set.seed(1729)
subcohort<-with(nwts, sample(1:nrow(nwts),750))
cases <- which(nwts$relaps==1)
nwts$in.cchsample<- (1:nrow(nwts)) %in% c(subcohort,cases)
nwts$in.subcohort<- (1:nrow(nwts)) %in% subcohort
nwts$wts<-ifelse(nwts$in.subcohort, 3915/750,1)
cch_design <- twophase(id=list(~1, ~1), subset=~in.cchsample,
    strata=list(NULL, ~in.subcohort), weights=list(NULL,~wts),
    data=nwts)
svycoxph(Surv(trel, relaps)~histol*stage+age+tumdiam,
    design=cch_design)

cch_data <- subset(nwts, in.cchsample)
cch_data$id <- 1:nrow(cch_data)
cch(Surv(trel, relaps) ~ histol*stage+age+tumdiam,  id=~id,
    data=cch_data, subcoh = ~in.subcohort, cohort.size=3915)
```

Figure 8.5 Classical case–cohort sampling from the NWTS population

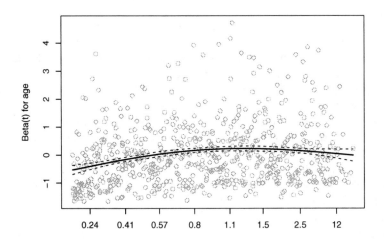

Figure 8.6 The assocation between age and hazard of relapse weakens over time

```
scch_design <- twophase(id=list(~1, ~1), subset=~in.cchsample,
    strata=list(NULL, ~relaps), data=nwts)
s1 <- svycoxph(Surv(trel, relaps)~histol*stage+age+tumdiam,
    design=scch_design)

nwts_design <- twophase(id=list(~1, ~1), subset=~in.subsample,
    strata=list(NULL, ~interaction(instit, relaps)), data=nwts)
s2 <- svycoxph(Surv(trel, relaps)~histol*stage+age+tumdiam,
    design=nwts_design)

v1 <- vcov(s1)
v2 <- vcov(s2)
diag(attr(v1, "phases")$phase2/attr(v1,"phases")$phase1)
diag(attr(v2, "phases")$phase2/attr(v2,"phases")$phase1)
```

Figure 8.7   Stratifed case–cohort sampling from the NWTS population, stratified on case status and stratified on both case status and local institution histology classification

Figure 8.7 shows code with a sample stratified on case status. The declaration of the design object is more straightforward here, as the weights can be computed by comparing the phase-one and phase-two samples. The stratified two-phase design analyzed with logistic regression in Figure 8.3 is also a stratified case–cohort design, and the same Cox model can be fitted, as shown.

Figure 8.7 also shows how to examine the contributions of variance from each stage of sampling. The variance matrix of a statistic computed from two-phase designs includes an attribute that gives the contributions of variance from each phase. In sampling from an existing cohort the first phase of sampling has been done in the past and cannot be changed. The phase-one contribution to variance estimates the irreducible minimum uncertainty that would remain even if everyone in the cohort were included in the second phase of sampling. The phase-two contribution is the remainder of the variance, due to including only a subsample of the available cohort in the analysis.

Table 8.6 shows the estimated log hazard ratios and standard errors for the two models in Figure 8.7 and for a Cox model using the complete data. The standard errors for the two case–cohort designs are divided into phase-one and phase-two contributions; the analysis of complete data has no phase-two component. The estimated phase-one standard errors in all three model fits are estimating the same quantity and should be very similar, as indeed they are.

In the two stratified designs about 1/3 of the participants are included in phase two. If these 1/3 were a simple random sample the total variance would be about three times the phase-one variance, so the phase-two contribution would be about twice the phase-one contribution. In fact, across the coefficients in the model the estimated phase-two contributions range from 1.1 to 1.7 times the phase-one contributions for the design stratified only on case status and 70 to 90% of the phase-one contributions

Table 8.6   Estimates from case–cohort designs stratified on case status alone, and on case status and local institution histologic classification, and from complete data from the NWTS population

	Stratified case–cohort		Full data
	relaps	relaps×instit	
Coefficient estimate			
histology	0.889	0.528	0.592
stage	0.196	0.256	0.221
age	0.050	0.024	0.056
tumor diameter	0.010	0.006	0.018
histology:stage	0.194	0.334	0.306
Standard error: (phase 1, phase 2)			
histology	(0.235, 0.287)	(0.234, 0.199)	0.237
stage	(0.043, 0.037)	(0.046, 0.041)	0.044
age	(0.016, 0.018)	(0.016, 0.014)	0.015
tumor diameter	(0.011, 0.012)	(0.012, 0.009)	0.010
histology:stage	(0.086, 0.114)	(0.086, 0.081)	0.084

for the design stratified on both case status and local institution histologic classification. Both designs are better than random subsampling, and stratifying on histologic classification provides real improvement over stratifying only on case status. A more efficient model-based analysis is known in theory (Nan [115]), but it is difficult to program and no software is currently available.

8.5   USING AUXILIARY INFORMATION FROM PHASE ONE

In a two-phase subsampling design there are, in principle, three ways that auxiliary information could be used. Auxiliary variables whose totals are known for the whole population can be used to reweight either the phase-one sample or the phase-two subsample, and variables measured at phase one can be used to reweight the phase-two subsample. There are typically few auxiliary variables available for the whole population, so we will focus on calibrating the second phase of sampling using auxiliary variables observed at phase one.

One important difference between this setting and Chapter 7 is that the the phase-one data set includes individual-level information on auxiliary variables, not just totals, expanding the range of possible reweighting techniques. Another is that, since the auxiliary information will be available to the data analyst, the reweighting can easily be customized to a particular analysis. A third difference is that, when sampling from an existing cohort, there may be a very large number of auxiliary variables available.

When estimating population means or totals, choosing auxiliary variables for two-phase designs follows much the same principles as in Chapter 7. For example, when estimating the prevalence of unfavorable histology in the case–cohort sample stratified only on `relaps`, there would be an increase in precision if the phase-two sample were calibrated using `instit`, which is correlated with histology.

As we saw in sections 7.3 and 7.4.1, it is more difficult to find auxiliary variables that increase the precision of subpopulation estimates or of regression parameter estimates. In particular, choosing variables that are correlated with the outcome variable or with predictor variables in a regression model is not very helpful. To see why this is the case and how to construct more effective auxiliary variables, we first go back to an artificial example with single-phase samples and population-level auxiliary variables, and then extend the lessons learnt to two-phase samples.

8.5.1   Population calibration for regression models

Consider the regression model fitted to a sample from the Academic Performance Index population in section 7.4.1. The auxiliary variable in that example was the 1999 Academic Performance Index, which is highly correlated with the outcome variable, the 2000 Academic Performance Index.

When estimating the mean 2000 API, population information on the 1999 API is helpful because a sample that, by chance, underestimates the mean 1999 API is likely to underestimate the mean 2000 API. When estimating the slope of a regression line against student mobility the information is not directly useful: underestimating the mean API for high values of student mobility leads to an underestimate of the slope, but underestimating the mean API for low values of student mobility leads to an overestimate of the slope. Whether the slope is overestimated or underestimated is almost independent of whether the mean of the outcome variable is overestimated or underestimated. For calibration to be useful in a regression model we need to construct a variable whose population mean or total approximates the estimated regression slope and then find auxiliary variables correlated with this constructed variable. The theory of *influence functions* shows how to do this. The description in this section is fairly loose and heuristic, more details can be found in Breslow et al. [16].

The influence function for an estimate $\hat{\beta}$ describes how the estimate changes when observations are added to or removed from the data. Write $\mathbb{I}(x_i, y_i; \beta)$ for the influence function evaluated at data (x_i, y_i) and parameter value β. Also write β^* for the true value of β, the value that would be obtained with complete population data. For nearly all estimates that are encountered in survey statistics, there is an explicitly computable influence function with the property

$$\hat{\beta} = \beta^* + \frac{1}{n} \sum_{i=1}^{n} \mathbb{I}(x_i, y_i; \beta^*) + \text{small remainder.}$$

Working out the influence function can be difficult for complex models, but the work has already been done for all the models and estimates we are using, since the linearization method of variance calculation requires these influence functions. In linear, generalized linear, and proportional hazards regression models the influence

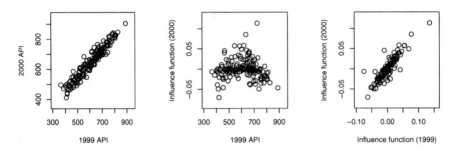

Figure 8.8   Correlations between Academic Performance Index for 1999 and 2000, and influence functions in regressions with 1999 and 2000 API as outcomes

functions evaluated at the observed data are just the $\Delta\beta$ deletion diagnostics (exactly or approximately, there are some slight variations on how these are defined).

Since any estimate can be approximated well by the population mean or total of its influence functions, good auxiliary variables for calibration will be highly correlated with these influence functions. The influence function $\mathbb{I}_{\hat{\beta}}$ for a linear regression slope in a model with a single predictor variable is

$$\mathbb{I}_{\hat{\beta}}(x_i, y_i; \beta) = \frac{1}{\pi_i} \frac{1}{\text{var}[X]} (x_i - \bar{x})(y_i - \mu_i(\beta)),$$

where $\mu_i(\beta)$ is the fitted value for individual i and $\bar{x}$ is the estimated population mean of X. If an auxiliary variable Z is highly correlated with Y it will have a low correlation with $\mathbb{I}_{\hat{\beta}}$, because the multiplier $(x_i - \bar{x})$ can be negative or positive with about equal probability.

If complete population data is available for Z and for X a better auxiliary variable can be constructed by fitting a linear regression of Z on X and taking the influence functions for this estimate. Figure 8.8 is based on the same example as in section 7.4.1. It shows the relationship between 1999 and 2000 Academic Peformance Index, between 1999 API and the influence function for the coefficient of `ell`, and between the influence functions for the coefficient of `ell` using 1999 API and using 2000 API as the outcome variable. The correlation between the two outcome variables is high, the correlation between the 2000 influence function and the 1999 outcome variable is low ($r = -0.05$), and the correlation between the two influence functions is high ($r = 0.88$).

Figure 8.9 shows code and results for calibration using 1999 API directly and using it through the influence functions. The influence functions are constructed by fitting a population model with 1999 API as the outcome, then using `dfbeta()` to extract the influence functions. The influence functions are then added as variables to the survey design object using `update()`, with `match()` used to work out which subset of the population collection of influence functions corresponds to the sample. The population totals for the influence functions are all zero, and `calibrate()` reweights the sample to make the sample totals also zero.

```
> m0 <- svyglm(api00~ell+mobility+emer, clus1)
>
> var_cal <- calibrate(clus1, formula=~api99+ell+mobility+emer,
      pop=c(6194,3914069, 141685, 106054, 70366),
      bounds=c(0.1,10))
> m1<-svyglm(api00~ell+mobility+emer, design=var_cal)
>
> popmodel <- glm(api99~ell+mobility+emer, data=apipop,
      na.action=na.exclude)
> inffun <- dfbeta(popmodel)
> index <- match(apiclus1$snum, apipop$snum)
> clus1if <- update(clus1, ifint = inffun[index,1],
      ifell=inffun[index,2], ifmobility=inffun[index,3],
      ifemer=inffun[index,4])
> if_cal <- calibrate(clus1if,
      formula=~ifint+ifell+ifmobility+ifemer,
      pop=c(6194,0,0,0,0))
> m2<-svyglm(api00~ell+mobility+emer, design=if_cal)
>
> coef(summary(m0))
              Estimate Std. Error   t value      Pr(>|t|)
(Intercept) 780.459500 30.0210123 25.997108 3.156974e-11
ell          -3.297892  0.4689026 -7.033215 2.173478e-05
mobility     -1.445370  0.7342887 -1.968395 7.473627e-02
emer         -1.814215  0.4233504 -4.285374 1.287085e-03
> coef(summary(m1))
              Estimate Std. Error   t value      Pr(>|t|)
(Intercept) 785.408240 13.7640081 57.062466 5.912274e-15
ell          -3.273108  0.6242978 -5.242864 2.756024e-04
mobility     -1.464732  0.6651257 -2.202188 4.989506e-02
emer         -1.676541  0.3742041 -4.480284 9.309647e-04
> coef(summary(m2))
              Estimate Std. Error    t value      Pr(>|t|)
(Intercept) 790.631553  5.8409844 135.359298 4.480786e-19
ell          -3.260976  0.1300765 -25.069679 4.678967e-11
mobility     -1.405554  0.2247022  -6.255187 6.214930e-05
emer         -2.240431  0.2150534 -10.418024 4.902863e-07
```

Figure 8.9   Calibration with population variables or population influence functions in the Academic Performance Index population

Calibration just using the variables `api99`, `ell`, `mobility`, and `emer` gives a substantial reduction in the intercept standard error, but has relatively little impact on the standard errors of the slope estimates. Calibration using the influence functions further reduces the standard error of the intercept and reduces the standard errors of all the slope parameters by a factor of 2–3.

8.5.2   Two-phase designs

In the example above we assumed that all variables but one were known for the whole population, and at an individual level, not just as totals. This is unreasonable for sampling from a population, but is a typical situation when taking a two-phase sample from an existing cohort, where only a few new variables will be measured on the subsample. In the examples cited earlier of recent case–cohort analyses using stored DNA or stored blood, each model being fitted included only one or at most a few phase-two variables.

Of more concern is whether the influence functions for the model of interest can be predicted effectively from phase-one information. When the phase-two variables are common genetic polymorphisms being screened for association with a phase-one phenotype it is unlikely that any phase-one variables will provide useful information about the genotype and calibration is unlikely to be helpful. When an interaction between a phase-one and phase-two variable is of interest, calibration may be helpful. The most promising situation, however, is when there is a phase-one variable that is strongly correlated with the phase-two variable.

Strong correlations are most likely to exist when a phase-one variable is a crude measurement of some quantity and the phase-two variable is a more accurate measurement. Self-reported smoking at phase-one could be followed up by urinary cotinine screening at phase two. "Has your doctor ever told you that you have high blood pressure?" could be followed up by blood pressure measurement and review of medications at phase two. A hospital discharge diagnosis of myocardial infarction, extracted from electronic records, could be followed up by review of detailed patient charts at phase two. Another related possibility is that the phase-two variable is associated with some broad biological category of variables, such as the acute-phase inflammatory response, and that other biomarkers in the same category measured at phase one would be correlated with it.

One reasonably general approach to constructing auxiliary variables based on influence functions is as follows [16]

1. Build an imputation model to predict the phase-two variable from the phase-one variables

2. Fit a model to all of phase one, using the imputed value for observations that are not in the phase-two sample

3. Use the influence functions from this model as auxiliary variables in calibration.

In the API example in the previous subsection the first step was simplified: we imputed the 2000 API by the 1999 API. Since the correlation between these variables was so high, a more sophisticated imputation model was unlikely to do any better.

Example: Wilms' tumor Breslow *et al.*[17] used the Wilms' tumor cohort data to illustrate this approach, in addition to reanalyzing data from a previously published case–cohort study in cardiovascular disease. The models fitted here are similar to those we used in that paper.

The first step is to construct an imputation model for the central lab histologic classification. This model is fitted to the phase-two sample, but uses only predictor variables from phase one. The most important predictor will be the local lab classification, but there may be additional information in other variables. The imputation model in Figure 8.10 follows Kulich and Lin [85]. The predicted values are extracted from the logistic regression model with `predict()`, into a variable `imphist` and the known central-lab histologic classification is used for the phase-two subsample where it is known.

The next step is to fit a model to the whole phase-one sample using the imputed histology variable, to provide influence functions that will be used as auxiliary variables. The influence functions for a Cox model are extracted using the `resid()` function. These influence functions are added to the `nwts` data set and a new survey object is defined. Finally, this new survey object is calibrated using the influence functions as auxiliary variables. The call to `calibrate()` specifies `phase=2`, indicating that the phase-two subsample is being calibrated to the phase-one sample. It is not necessary to specify totals for the auxiliary variables as the full phase-one data are stored in the survey design object and these totals can readily be computed. Raking calibration is used to ensure non-negative weights, which the Cox regression functions require. The first two models, `m1` and `m2`, are fitted to the two-phase sample using sampling weights and calibrated weights, respectively.

Since this procedure requires imputing the central-lab histologic classification for all children in the study, another approach would be to use this imputed value directly as a predictor and fit an unweighted model to the whole sample (`m3`). This would be a standard choice if the analysis were viewed as a measurement-error problem. The coefficient estimates are slightly biased even if the imputation model is fits well, though not enough to cause practical problems in interpretation. The bias can be more serious when the imputation model fits poorly.

Table 8.7 shows the coefficient estimates and standard errors from these three models and model `m4` that uses the complete data on central lab histology. Both raking and imputation reduce the standard errors. This is especially true for variables that are observed for everyone in the sample, where the standard errors are close to those that would be obtained with complete data. The results are very similar for raking and direct imputation, which tends to be the case when the imputation model is good. The raking approach has the additional advantage of always being valid, without any assumptions about models, and of always being at least as accurate as the two-phase analysis based on sampling weights.

```
impmodel <- glm(histol~instit+I(age>10)+I(stage==4)*study,
  data=nwts, subset=in.subsample, family=binomial)
nwts$imphist <- predict(impmodel, newdata=nwts, type="response")
nwts$imphist[nwts$in.subsample] <- nwts$histol[nwts$in.subsample]

ifmodel <- coxph(Surv(trel,relaps)~imphist*age+I(stage>2)*tumdiam,
  data=nwts)
inffun <- resid(ifmodel, "dfbeta")
colnames(inffun) <- paste("if",1:6,sep="")

nwts_if <- cbind(nwts, inffun)
if_design <- twophase(id = list(~1, ~1), subset = ~in.subsample,
  strata = list(NULL, ~interaction(instit, relaps)),
  data = nwts_if)
if_cal <- calibrate(if_design, phase=2, calfun="raking"
  ~if1+if2+if3+if4+if5+if6+relaps*instit)

m1 <- svycoxph(Surv(trel, relaps)~histol*age+I(stage>2)*tumdiam,
  design=nwts_design)
m2 <- svycoxph(Surv(trel, relaps)~histol*age+I(stage>2)*tumdiam,
  design=if_cal)
m3 <- coxph(Surv(trel, relaps)~imphist*age+I(stage>2)*tumdiam,
  data=nwts)
m4 <- coxph(Surv(trel, relaps)~histol*age+I(stage>2)*tumdiam,
  data=nwts)
```

Figure 8.10   Calibrating to influence functions based on imputed histologic classification

Table 8.7   Coefficient and standard errors for Cox models without calibration (m1), calibrated to influence functions (m2), using imputed histology directly(m3), and using the complete data(m4)

	Two-phase sample			Full data
	sampling weights	raked	direct imputation	
Coefficient estimate				
histology	1.808	2.113	2.108	1.932
age	0.055	0.101	0.101	0.096
stage > 2	1.411	1.435	1.432	1.389
tumor diameter	0.043	0.061	0.061	0.058
histology:age	-0.116	-0.159	-0.159	-0.144
stage > 2:diametee	-0.074	-0.084	-0.083	-0.079
Standard error				
histology	0.221	0.171	0.174	0.157
age	0.023	0.014	0.016	0.016
stage > 2	0.361	0.276	0.249	0.250
tumor diameter	0.021	0.016	0.014	0.014
histology:age	0.054	0.039	0.040	0.035
stage > 2:diameter	0.030	0.022	0.020	0.020

On the other hand, in small phase-two samples or with low event rates direct imputation has the advantage of not requiring any events to occur in the phase-two subsample. For example, the Women's Health Initiative trial measured protein and energy intake in using biomarkers in a small subsample of participants chosen at the start of the study [118]. Only a handful of heart disease or cancer events would be expected in this subsample and a two-phase sampling approach would be ineffective.

8.5.3   Some history of the two-phase calibration estimator

Särndal and Swensson [150] reviewed two-phase estimation, including regression estimators for the population total using either population or phase-one auxiliary variables, but they did not consider estimating other statistics. Robins, Rotnitzky, and Zhao [137] characterized all the valid estimators for regression coefficients in a two-phase design and showed how (in theory) the most efficient design-based and model-based estimators were constructed. Their "Augmented Inverse-Probability Weighted" (AIPW) estimators are related to design-based calibration estimators, and the most efficient AIPW estimators corresponds to the calibration estimator with the optimal set of auxiliary variables. In practice it is not feasible to construct this estimator, since it involves expected values taken over an unknown population distribution.

Breslow and Chatterjee [13] and Borgan et al. [11] described post-stratification for regression models in two-phase samples, for logistic regression and Cox regression respectively. Kulich and Lin[85] constructed a complicated ratio-type estimator based on adjusting the sampling weights separately for each time point and each predictor variable. The strategy of imputing the phase-two variable and using influence functions as auxiliary variables is a simpler version of their approach. Breslow and Lumley and coworkers [16, 17] describe applications and theory for the approach based on influence functions. Mark and Katki [103] followed the approach of Robins and colleagues more directly and worked out the design-based estimator that would be optimal under particular parametric models. This is implemented it in the R package NestedCohort [71]. NestedCohort has the advantage of being able to compute standard errors for survival curves, which survey did not implement at the time of writing. All these approaches give similar results for examples based on the Wilms' tumor population; it is not clear how they compare in general. Until very recently there has been no overlap between the survey literature based on calibration or GREG estimation and the biostatistics literature descended from Robins, Rotnitzky, and Zhao; as of mid-2008 the Web of Science citation system listed no paper that cited both [137] and one of [40, 41, 150].

EXERCISES

8.1 ⋆ Suppose a phase-one simple random sample of size n is taken from a population of size N, to measure a variable X with G categories. Write $N_1, N_2, \ldots, N_G$ for the (unknown) number of individual in the population in each category, and $n_1, n_2, \ldots, n_g$ for the number in the phase-one sample. The phase-two sample takes a fixed number m from each category. Show that π_i^* and π_{ij}^* for this design approach π_i and π_{ij} for a stratified sample from the population as n increases.

8.2 Construct a full two-phase data set for the Ille-et-Vilaine case–control study. The additional phase-one observations are $430000 - 975$ controls to make the number up to the population size. Fit the logistic regression model using the design produced by twophase() and compare the results to the weighted estimates in Figure 8.1.

8.3 This exercise uses the Washington State crime data for 2004 as the population. The data consist of crime rates and population size for the police districts (in cities/towns) and sheriffs' offices (in unincorporated areas), grouped by county

    a) Take a simple random sample of 20 police districts from the state and use all the data from the sampled counties. Estimate the total number of murders and of burglaries in the state

    b) Calibrate the sample in the previous question using population as the auxiliary variable, and estimate the total number of murders and of burglaries in the state

    c) Take a simple random sample of 100 police districts as phase one of a two-phase design, and assume that population is the only variable available at phase one. Divide the sample into 10 strata with roughly equal total population and sample two police districts from each stratum for phase two. Estimate the total number of murders and of burglaries in the state.

    d) Calibrate the two-phase sample using the phase-one population size data. Estimate the total number of murders and of burglaries in the state.

8.4 ⋆ The sampling probabilities π and π^* in the NWTS two-phase case–control study depend on the 2 × 2 table of `relaps` and `instit`. Suppose that the superpopulation probabilities for the cells in the 2 × 2 table match those in Table 8.3

　　a) Write R code to simulate realizations of Table 8.3 and to compute the second-phase sampling probabilities $\pi_{i(2|1)}$ for a two-phase design with sample size 1300 and cell counts as equal as possible. That is, sample everyone in a cell that has fewer than 1300 people and then divide the remaining sample size evenly over the remaining cells.

　　b) Run 1000 simulations, compute π_i as the average of π_i^* for a given cell over the simulations, and compare π_i to the distribution of π_i^*.

8.5 This exercise uses the Washington State crime data for 2004 as the population and the data for 2003 as auxiliary variables.

　　a) Take a simple random sample of 20 police districts from the state and use all the data from the sampled counties. Estimate the ratio of violent crimes to property crimes.

　　b) Calibrate the sample in the previous question using population and number of burglaries as the auxiliary variables, and estimate the ratio of violent crimes to property crimes

　　c) The ratio of violent crimes to property crimes in the state in 2003 was 21078/290945=0.0724. Define an auxiliary variable `infl = violent −` 0.0724 × `property`, the influence function for the ratio, and use it to calibrate to the state data. Estimate the ratio of violent crimes to property crimes

　　d) Take a simple random sample of 100 police districts as phase one of a two-phase design, and assume that the 2003 crime data are available at phase one. Divide the sample into 10 strata with roughly equal total number of burglaries in 2003 and sample two police districts from each stratum for phase two. Estimate the total number of murders, and of burglaries, in the state.

　　e) Calibrate the two-phase sample using the auxiliary variable `infl`. Estimate the total number of murders, and of burglaries, in the state.

8.6 What would the efficiency approximation from exercise 5.8 give as the loss of efficiency from using weights in a case–control design?

CHAPTER 9

MISSING DATA

In which we take what we are given.

9.1 ITEM NON-RESPONSE

This chapter deals primarily with *item non-response*, where partial data are available for a respondent. As we saw in Chapter 7, the bias from *unit non-response*, or failure to recruit a sampled individual, can be mitigated by adjusting the sampling weights, thus modelling the non-response as part of the designed sampling mechanism. For item non-response the situation is more favorable, but more complex, because the partial information available on the participant can be used to improve estimation.

There are two broad classes of approach to item non-response. One is to model the non-response as part of the sampling mechanism, in a two-phase design in which some variables are measured on the whole sample and others on a subsample. The other is to impute the missing data, using the observed information on each subject as a guide to plausible values for the missing information. Multiple imputation and reweighting are sometimes described as "statistically principled" approaches to

Complex Surveys: A Guide to Analysis Using R. By Thomas Lumley
Copyright © 2010 John Wiley & Sons, Inc.

inference with missing data. This means that there are well-characterized sets of assumptions under which these approaches will give correct answers.

The necessary assumptions are versions of Rubin's [141] *Missing At Random* condition, stating that all outcome differences between responders and non-responders are explained by the completely observed auxiliary variables. Unfortunately, these sets of assumptions are *a priori* implausible in many applications — non-responders really are different — and cannot be checked using the observed data. The situation is not completely hopeless, however. Using multiple imputation or reweighting to address non-response is not likely to remove bias completely, but it is reasonable to expect bias to be reduced when compared with an analysis that simply drops the missing observations.

9.2   TWO-PHASE ESTIMATION FOR MISSING DATA

In a probability sample with item non-response, some variables will be observed on everyone in the sample and some will be observed only on a subset. We can think about this as the result of a two-phase design: the original sample is selected and the variables with no missing data are measured, then a subset of people is selected and the remainder of the variables are observed (we write $R_i = 1$ for an individual in the subset and $R_i = 0$ otherwise). If the subset had been chosen as a probability sample the result would be a two-phase probability sample and could be analyzed using the methods of Chapter 8. In reality, the subset of individuals with complete data is not a probability sample, but pretending the data are a two-phase probability sample is still a useful strategy. The aim in treating the data as a two-phase sample is to use the variables with complete data as auxiliary information to mitigate the bias from non-response.

A further complication is that there may not only be two sets of individuals to consider. In an analysis involving only two variables X and Y, there could be observations with both X and Y observed, with X but not Y, and with Y but not X. If the analysis is a regression model with several predictor variables there could be hundreds of subsets. The two-phase approach will only be useful when it is possible to divide the observations into two sets, complete and incomplete. In the two-variable example, if there are many observations with X but not Y and few with Y but not X, it might be possible to omit the observations with Y but not X, considering them as having unit non-reponse.

9.2.1   Calibration for item non-response

A calibration approach to using the complete variables starts by defining the phase-two sampling probabilities $\pi_{i(2|1)}$ simply as the proportion of complete observations, so that the two-phase sampling weights $1/\pi_i^*$ are proportional to the sampling weights $1/\pi_i$ for the original design. Calibration is then used to adjust the weights, correcting for differences between the complete and incomplete observations, and the calibrated weights g_i/π_i^* are used in analysis.

```
expit <- function(eta) exp(eta)/(1+exp(eta))

pmar <- with(apistrat,expit(-7+api99/100-emer/10))
pnar <- with(apistrat,expit(-7+api00/100-emer/10))
mar<-rbinom(nrow(apistrat), 1, pmar)
nar<-rbinom(nrow(apistrat), 1, pnar)

stratmar <- apistrat
stratmar$api00[mar==1] <- NA
stratmar$w2 <- nrow(apistrat)/sum(1-mar)
stratnar <- apistrat
stratnar$api00[nar==1] <- NA
stratnar$w2 <- nrow(apistrat)/sum(1-nar)

mar_des <- twophase(id=list(~1,~1), strata=list(~stype,~stype),
    subset=~I(!is.na(api00)), weights=list(~pw,~w2 ),data=stratmar)
nar_des <- twophase(id=list(~1,~1), strata=list(~stype,~stype),
    subset=~I(!is.na(api00)), weights=list(~pw,~w2 ),data=stratnar)

calmar1 <- calibrate(mar_des, phase = 2, calfun = "raking",
    ~api99 + emer + stype + enroll)
calnar1 <- calibrate(nar_des, phase = 2, calfun = "raking",
    ~api99 + emer + stype + enroll)
calmar2 <- calibrate(mar_des, phase = 2, calfun = "raking",
    ~ns(api99,3) + emer + stype + enroll)
calnar2 <- calibrate(nar_des, phase = 2, calfun = "raking",
    ~ns(api99,3) + emer + stype + enroll)

dstrat<-svydesign(id = ~1, strata = ~stype, weights = ~pw,
    data = apistrat, fpc = ~fpc)
svymean(~api99+api00,dstrat)
svyglm(api00~emer+ell+meals, dstrat)
```

Figure 9.1 Simulating missingness and calibration in the Academic Performance Index population. MAR is missing at random, NAR is missing not at random, depending on the unknown api00 value.

Table 9.1   Percentage bias in means and in regression coefficients in a model for 2000 API, in the missing data scenarios in Figure 9.1 (based on 1000 simulations). Scenario MAR is missing at random, NAR is missing not at random.

| | MAR | | | NAR | | |
	raw	linear	flexible	raw	linear	flexible
Means						
api99	-7	0	-5	-7	0	-5
api00	-6	0	-4	-7	0	-5
Coefficients						
(Intercept)	-4	-1	-2	-5	-2	-2
emer	-15	-20	12	-19	-22	12
ell	31	15	25	25	7	19
meals	-16	-7	-8	-17	-7	-9

Figure 9.1 shows a simulation examining the effects of calibration in the Academic Performance Index population. The intended sample is a stratified sample of schools, in the built-in data set `apistrat`. Missingness probabilities for the 2000 API variable are defined based on proportion of teachers with emergency qualificaitons (`emer`) and either the 1999 or 2000 API variables. The `expit` function used to define the probabilities is the inverse of the logit function and the `rbinom()` function generates a 0 or 1 with the specified probability.

When missingness depends on the 2000 API variable itself, the Missing At Random assumption does not hold; when missingness depends on the completely observed 1999 API and `emer` the assumption does hold. Since the 1999 and 2000 API variables are very strongly correlated, it would not be surprising to see similar behavior from the two missingness scenarios despite the important technical difference. A two-phase design object is defined for each data set, with the phase-two sampling probabilities simply being the proportion that have observed data for `api00`. The two design objects are calibrated to four complete-data variables: 1999 API, proportion of teachers with emergency qualifications, school type, and school size. The first two are directly related to the way non-response has been generated, the last two are variables that might be expected to affect response rates in real life. A second set of calibrated designs uses a more flexible, non-linear model for `api99`, a cubic spline with three degrees of freedom. Finally, the survey design object `dstrat` contains the whole sample, without any missing data.

Table 9.1 shows the results of two analyses run on these survey design objects. The first analysis estimates the population mean for 1999 and 2000 API. The second analysis is a linear regression of 2000 API on three socio-economic variables. These analyses are run without any adjustment for missingness ("raw"), using the simple calibration model ("linear") and using the more flexible calibration model ("nonlinear") and the whole process is repeated 1000 times. The table shows the percentage bias, relative to the sample with no missing data.

Without any adjustment for missingness the bias is large; the simulation makes response depend strongly on Academic Performance Index. Calibration produces a substantial reduction in the bias, essentially eliminating bias for the two means and the regression intercept and reducing it by half or more for two of the regression coefficients. The more flexible calibration model performs worse than the simple model. This should be surprising: making a calibration model more flexible should match the subsample to the complete data more effectively. The problem is the relatively small sample size, about 150 complete and 50 incomplete observations. The number of observation with high values for `api99` is already small, and fitting a more flexible model allows the noise from chance variations in non-response to outweigh the potential extra accuracy. None of the analyses gives an accurate estimate for the regression coefficient of `emer`. Again, this is due in part to the small sample size and the relatively strong correlation between API and `emer`, which makes it difficult to separate their effects on missingness.

9.2.2   Models for response probability

An alternative to post-stratification and calibration is to model the phase-two response probabilities $\pi_{i(2|1)}$ using logistic regression (or any other binary regression model). Logistic regression has the practical advantage over calibration of always giving probabilities between 0 and 1. It may also be easier to understand the description of the missingness process that arises from a logistic regression model. To avoid confusion with regression coefficients that might be being estimated in the analysis, we will use γ for the coefficients in the logistic model for missingness:

$$\text{logit Pr}[R_i = 1] = \text{logit}\,\pi_{i(2|1)} = x_i\gamma.$$

The calibration and logistic regression approaches are closely related. The weights that result from logistic regression do not (in general) satisfy the calibration constraints

$$\sum_{i=1}^{n} \frac{1}{\pi_{i(2|1)}} x_i = \sum_{j=1}^{N} x_j \qquad (9.1)$$

but they do satisfy a similar set of constraints

$$\sum_{i=1}^{n} x_i = \sum_{j=1}^{N} \pi_{j(2|1)} x_j \qquad (9.2)$$

where now the probabilities are on the right-hand (population) side of the equation, rather than the left-hand (sample) side. The calibration constraints (9.1) say that the estimated phase-one total must match the actual phase-one total. The logistic regression constraints (9.2) say that the observed phase-two total must match the expected phase-two total. When the auxiliary variables x are a complete cross-classification of a set of categorical vectors the two sets of constraints are identical, so that post-stratification can be viewed either as a special case of estimating the sampling probabilities or as a special case of calibration.

Example: NHANES III bone density. The difficulty with real examples of missing data is that the truth is not known. As an intermediate case we can compare calibration for missing data in NHANES III to the estimates from multiple imputation that will be described in more detail in section 9.3.2. We will examine data on bone mineral density, using the code in Figure 9.2.

The data are stored in a SQLite database and need to be loaded into R because analysing data directly from the database is not supported for two-phase designs. The call to dbGetQuery() loads the necessary variables. Of the roughly 30,000 individuals in the NHANES III sample the bone mineral density is "not applicable" for 15,169, measured for 14,646, and missing for 4179, as indicated by the imputation flag variable BDPFNIF. The individuals where bone mineral density is not applicable are removed, and the variable hasbone is defined as 1 for observed and 0 for missing. A logistic regression model for this variable uses weight, age, sex, race/ethnicity, region of the country, and whether the participant lives in a rural or urban region.

Three survey design objects are defined. The first, design0, has a constant phase-two sampling probability, corresponding to a constant non-response probability. Using this design ignores any non-response bias. The second, design1, uses fitted sampling probabilities from the logistic regression model. The third, design2, uses the same auxiliary variables via calibration.

The mean bone mineral density estimated from design0 is 0.8256 g/cm^{-2} with standard error 0.003. Modelling the response probabilities gives 0.8185, with standard error 0.003, and calibration gives 0.8171, with standard error 0.003. Although the impact of the non-response adjustment appears small, it is larger than the standard error of the estimate and so has potential to affect inference. The difference between the adjusted and unadjusted analyses suggests that bone mineral density was lower in people who did not have it measured than in those who did. The same conclusion would be reached from the multiply imputed data, where the mean of the imputed values is 0.809 and of the observed values is 0.827.

In large enough samples these reweighting methods will not introduce bias if none already exists; if non-response is already independent of the variables being analyzed then the reweighting will simply increase precision. This does not mean that calibration always results in more accurate inference. Consider a survey that estimates mean income, in which women are less likely to respond to the income question than men with the same income, and where people with higher income are less likely to respond than those with lower income. As men have higher average income than women, the two sources of non-response bias tend to cancel. Calibration on gender would remove the gender bias, unmasking the high-income bias and making the estimate of mean income less accurate.

9.2.3 Effect on precision

The overall impact of either calibration or estimating response probabilities on precision is hard to predict. To see why, imagine a situation where the true response probabilities $\pi_{i(2|1)}$ are known and also a completely accurate model is available to estimate them, with no problems of insufficient sample size. Going from con-

```
library(RSQLite)
sqlite<-dbDriver("SQLite")
dbcon<-dbConnect(sqlite, dbname="~/nhanes/imp.db")
nhanes <- dbGetQuery(dbcon,"select SDPPSU6, WTPFQX6, SDPSTRA6,
    HSAGEU,   HSAGEIR,DMARETHN,HSSEX,DMARETHN, DMPMETRO,
    DMPCREGN , BMPWTMI,BMPWTIF, BDPFNDMI,BDPFNDIF from set1")

nhanes <- subset(nhanes, BDPFNDIF>0)
nhanes$hasbone <- 2-nhanes$BDPFNDIF
nhanes$age <- with(nhanes,
    ifelse(HSAGEU==1, HSAGEIR/12, HSAGEIR))
nhanes <- subset(nhanes, age>20)
nhanes$psu <- with(nhanes, SDPPSU6+10*SDPSTRA6)
nhanes$agegp <- with(nhanes, cut(age, c(20,40,60,Inf)))

model<-glm(hasbone~(BMPWTMI+age)*(HSSEX*DMARETHN+
    DMPMETRO*DMPCREGN), family=binomial, data=nhanes)

nhanes$p0<-mean(nhanes$hasbone)
nhanes$pi2<-fitted(model)

design0<-twophase(id=list(~psu, ~1),strata=list(~SDPSTRA6, NULL),
    weights=list(~WTPFQX6, ~I(1/p0)), subset=~I(hasbone==1),
    data=nhanes)
design1<-twophase(id=list(~psu, ~1),strata=list(~SDPSTRA6, NULL),
    weights=list(~WTPFQX6, ~I(1/pi2)), subset=~I(hasbone==1),
    data=nhanes)
design2<-calibrate(design0, phase=2,
   formula=~(BMPWTMI+age)*(HSSEX*DMARETHN+
      DMPMETRO*DMPCREGN))
```

Figure 9.2   Calibration for missing bone mineral density data in NHANES III

stant phase-two sampling probabilities based on the overall proportion of response to the true response probabilities will remove the non-response bias, but will tend to increase the variance of the estimate because the weights become more variable. Going from the true response probabilities to the model-estimated response probabilities will have no impact on bias but will decrease the variance of the estimate, because the observed-data sample is better matched to the complete sample.

Potential auxiliary variables for any particular analysis can be divided into three groups:

1. associated with non-response but not with the variables being analyzed,

2. associated with non-response and with the variables being analyzed,

3. associated only with the variables being analyzed and not with non-response.

Auxiliary variables in the third category will increase precision, with no effect on bias. Those in the second category will reduce bias, and may increase or decrease precision as explained above. Auxiliary variables in the first category will increase the variability of weights and decrease precision, but without any reduction in bias.

Ideally, an analysis would use only the latter two categories of variables. The problem is that it is usually impossible to be confident that a potential auxiliary variable is really independent of all the variables being analyzed, so it is prudent to include any auxiliary variable that strongly predicts non-response, regardless of its apparent association with other variables. This is especially true when computing a set of weights that will be used for many future analyses of many different sets of variables. An additional issue when designing weighting schemes for public use data is that users may not be able to compute standard error estimates that take into account the precision gains from calibration, in which case reduction of bias is the only relevant goal.

9.2.4 ★ Doubly-robust estimators

We have seen that reweighting estimators for the population total of Y can constructed as a two-phase estimators based on estimated response probabilities

$$\hat{T}_{est} = \sum_{i=1}^{n} \frac{R_i}{\pi_i^*} Y_i$$

or as calibration estimators

$$\hat{T}_{cal} = \sum_{i=1}^{n} \frac{g_i R_i}{\pi_i} Y_i$$

correcting for the bias that results from using the intended sampling probabilities π_i. As we saw in Chapter 7, the calibration estimator is just a regression estimator for a working model $E[Y|X = x] = \mu(x, \hat{\beta})$

$$\hat{T}_{cal} = \sum_{i=1}^{n} \frac{R_i}{\pi_i p} (Y_i - \mu_i(x_i, \hat{\beta})) + \sum_{i=1}^{n} \frac{1}{\pi_i} \mu_i(x_i, \hat{\beta}),$$

where p is the overall proportion of non-response.

The two-phase estimator will be valid if the model for the response probabilities correctly models everything that affects both Y and non-response. The calibration estimator will be valid if the working model Y correctly models everything that affects both Y and non-response.

When the reweighting is simply post-stratification, based on dividing the sample into homogenous groups, the assumptions needed for the response model to be correct and for the working model for Y to be correct are the same: that non-response is independent of Y within post-strata. For more complex calibration estimators the assumptions are not the same, for example, a raking estimator could be unbiased because the mean of Y is an additive function of x or because the response probability is a multiplicative function of x, and these are distinct possibilities.

Rather than having the model for response probability implied by a working model for the mean or *vice versa*, it is possible to have separate models for $\pi_{i(2|1)} = P[R = 1|X = x]$ and for $\mu(x, \beta) = E[Y|X = x]$, so that the estimator will be unbiased in large samples if either model is correct, and will have the added efficiency of a calibration estimator if both models are correct. This *double robustness* property is due to Robins and Rotnitzky [136].

Double robustness is useful only when the set of available auxiliary variables x is large. If there are few xs it is possible to calibrate on all of them and the only reason for residual nonresponse bias is that the auxiliary variables are insufficient. If there are many auxiliary variables then it is not possible to use all of them in calibration. Model selection will be necessary, and selecting a sufficiently good model from a large set of variables requires substantive knowledge. The advantage of a double-robust estimator is that there are two opportunities to inject substantive knowledge: the non-response mechanism and the model for Y.

For more general statistics than an estimated population mean it may be difficult or impossible to construct a double-robust estimate, because the mathematical theory assumes the non-response model and the regression calibration model are mathematically compatible; that they could both be exactly correct. It is still reasonable to expect some benefit in model robustness from fitting both a non-response using substantive knowledge about reasons for non-response and a calibration model using knowledge about relationships between outcomes. Robins and Rotnitzky [136] call these *generalized double robust* estimates.

9.3 IMPUTATION OF MISSING DATA

Imputation involves filling in missing data to produce a complete data set. This is obviously convenient for data analysis, as any analysis that can be done with complete data can also be done with imputed data. Imputation is especially attractive in public-use data sets, where it is not possible to predict what analyses will be needed in the future. The imputed values can be taken from other individuals in the sample, either at random or matched on key variables (*hot-deck imputation*), or they can be produced from a parametric model fitted to the complete data.

If a relationship between variables is not incorporated in the imputation process, that relationship will not be present in the imputed data. There are two ways that this can create bias. Suppose that health insurance status is being imputed, and that type of employment is not being used to create the imputed values, but type of employment actually does affect health insurance status. One source of bias is that type of employment may affect the probability of non-response: unemployed people may be easier to contact, those working multiple part-time jobs may be harder to contact. Failing to use this information will typically mean that the population distribution for insurance status is incorrect and that all estimates will be biased.

Even if employment type did not affect non-response, or if the dependence had already been removed by raking, imputations that ignore employment type can give correct population distributions. There will still be bias when trying to estimate the association between employment type and imputed variables. As the imputed data are imputed without an association between employment type and insurance status, the estimated association will be weaker than the true association in the population.

Carefully constructed imputations can be valuable despite these problems, and are quite widely used. The California Health Interview Survey constructs imputed values for all measurements except for "some sensitive variables in which non-response had its own meaning" [24]. CHIS used a hot-deck imputation method that chooses a value from another individual based on a set of matching variables, with less important matching variables dropped where necessary. These matching variables included gender, age, race/ethnicity, poverty level, education, and region, for all imputed variables and other matching variables as appropriate to the variable being imputed. Only about 1% of the data was imputed, so the impact on bias and variance is likely to be small, but the benefit for analytical convenience is considerable. Flags for imputed data and the original type of non-response are also provided: for example, the question "Has a doctor ever told you that you have cancer?" had 51 "Don't Know" and 22 "Refused" non-responses, which were imputed for 8/51 and 3/19 as "Yes."

Any form of imputation that produces a single complete data set has the disadvantage that the uncertainty in the missing data is not represented in the imputed data set. When the amount of missing data is more than a few percent this uncertainty can be important. In multiple imputation, developed by Rubin [143, 144, 145], multiple complete data sets are constructed, so the uncertainty in imputation is reflected by the differences between the data sets.

The first step in multiple imputation is to construct a model that can be used to predict the missing data, and fit this model to the observed data. Once this model is constructed, the missing data are sampled from the predictive distribution of the model. For example, if the model were a linear regression model with Normal residuals, the imputations would be created as the fitted mean from the model plus a random Normal error. The imputation procedure is performed multiple times, and the result is a set of plausible complete data sets. For computational reasons the number of complete data sets m is often fairly small, perhaps $m = 5$–10. As with raking and calibration for unit non-response, the modeling and imputation procedure need only be done once, and can take advantage of variables that will not be included in public-use data sets.

At the analysis stage, the same analysis is performed on each of the imputed data sets, to produce a set of estimates $\hat{T}_1, \hat{T}_2, \ldots, \hat{T}_m$ and estimated variances $v_1, v_2, \ldots,$ v_m. If the imputation model correctly reflects all the relationships in the data then each of the estimates $\hat{T}_1, \ldots, \hat{T}_m$ will be approximately unbiased, so a good overall estimate is the average

$$\bar{T} = \frac{1}{m} \sum_{j=1}^{m} \hat{T}_j. \tag{9.3}$$

The variances based on the complete data sets will be too small: they will be approximately unbiased estimates of the variance if there were no non-response. Rubin proposes that the additional variance due to non-response is estimated by the variance between the m estimates $\hat{T}_1, \ldots, \hat{T}_m$, which differ only because of the uncertainty in the missing data

$$\widehat{\text{var}}\left[\bar{T}\right] = \frac{1}{m} \sum_{j=1}^{m} v_j + \left(\frac{m+1}{m}\right) \frac{1}{m-1} \sum_{j=1}^{m} \left(\hat{T}_j - \bar{T}\right)^2. \tag{9.4}$$

This variance formula is strictly valid only for so-called *proper imputation*, where the imputations are created from the posterior distribution of a parametric Bayesian model. In addition, the imputation model, because it makes predictions using the entire data set, can give more precise estimates for small subpopulations than would be available from a strictly design-based estimator that did not use information from outside the subpopulation [74]. Appendix A.4 gives more details.

Constructing good imputation models requires specialist knowledge about the data and uses statistical and computational methods that are outside the scope of this book. Using multiply imputed data, on the other hand, is a straightforward part of survey analysis.

9.3.1   Describing multiple imputations to R

Analysis of multiply imputed data uses the mitools package [100] in addition to the survey package. There are three steps in analysis:

1. create an object that contains a survey design for each imputed data set

2. compute an estimate on each of the data sets

3. combine the estimates to get a single estimate and standard error.

The separate imputed data sets can be loaded into R, or can be stored as tables in a relational database. The function `imputationList()` creates an object containing a list of data sets. When this object is supplied as the `data=` argument to `svydesign()` the result is a list of imputed survey designs.

In the second step, the function `with()` can be used to run any survey analysis on each of the imputed data sets and return a list of results.

Finally, the function `MIcombine()` combines the repeated data sets using equations 9.3 and 9.4, and reports the results.

```
> library(mitools)
> library(RSQLite)
> impdata <- imputationList(c("set1","set2","set3","set4","set5"),
           dbtype="SQLite", dbname="~/nhanes/imp.db")
> impdata
MI data with 5 datasets
Call: imputationList(c("set1", "set2", "set3", "set4", "set5"),
    dbtype = "SQLite", dbname = "~/nhanes/imp.db")
> designs <- svydesign(id=~SDPPSU6, strat=~SDPSTRA6,
    weight=~WTPFQX6, data=impdata, nest=TRUE)
> designs
DB-backed Multiple (5) imputations: svydesign(id = ~SDPPSU6,
    strat = ~SDPSTRA6, weight = ~WTPFQX6,
    data = impdata, nest = TRUE)
```

Figure 9.3   Describing the NHANES III imputation data to R

9.3.2 Example: NHANES III imputations

The NHANES III imputation project created five multiple-imputation data sets for approximately 60 variables. The imputation procedures are described by Schafer [154] and the data are available as NHANES III data set 7a, together with a set of sample analyses [153].

The data are supplied as fixed-format text in six files: a core file of variables that are not imputed, and five imputation files. Section D describes reading these data sets into the SQLite relational database and merging each imputation file with the core file.

The first analysis step is to create the list of survey designs, as shown in Figure 9.3. The first argument to imputationList() is the data sets. When the data sets are stored in database tables they are supplied as the names of these tables; if they were data sets loaded into R the first argument would be list(set1,set2,set3, set4, set5). The imputation list is passed as the data argument to svydesign(), which created a list of survey designs. When data are kept in a database rather than loaded into R it is necessary that the strata, clusters, and sampling weights are the same for all the imputations.

The first example analysis given by Schafer [153] is a tabulation of means for a set of variable by age and sex. Figure 9.4 shows this analysis for one variable, bone mineral density in the neck of the femur (BDPFNDMI). The first step is to create an age group variable using update() by combining the two supplied age variables and using cut() to define categories. The call to with() runs a svyby() call on each of the five imputed data sets and res is a list of five sets of results. Finally, MIcombine() combines these five data sets to give the estimates, standard errors, and confidence intervals.

```
> designs<-update(designs,
    age=ifelse(HSAGEU==1, HSAGEIR/12, HSAGEIR))
> designs<-update(designs,
    agegp=cut(age,c(20,40,60,Inf),right=FALSE))
> res <- with(subset(designs, age>=20),
    svyby(~BDPFNDMI, ~agegp+HSSEX, svymean))
> summary(MIcombine(res))
Multiple imputation results:
    with(subset(designs, age >= 20), svyby(~BDPFNDMI,
        ~agegp + HSSEX, svymean, design = .design))
    MIcombine.default(res)
               results         se    (lower     upper) missInfo
[20,40).1  0.9355049 0.003791945 0.9279172 0.9430926     28 %
[40,60).1  0.8400738 0.003813224 0.8325802 0.8475674     10 %
[60,Inf).1 0.7679224 0.004134875 0.7598032 0.7760416      8 %
[20,40).2  0.8531107 0.003158246 0.8468138 0.8594077     26 %
[40,60).2  0.7839377 0.003469386 0.7771144 0.7907610     11 %
[60,Inf).2 0.6454393 0.004117235 0.6370690 0.6538096     38 %
```

Figure 9.4   Bone mineral density by age and sex, from NHANES III imputation data

An estimate of the percentage of missing information is also reported. This is approximately the between-imputations variance as a fraction of the total variance. It is usually lower than the proportion of observations that are missing, reflecting the information gained about those variables from the observed variables. The percentage of missing information can be higher than the proportion of observations missing when the missing observations are particularly influential. The proportion of missing information is similar to the design effect, in that it compares the precision of an estimate to the precision that would be available in a study of the same size without non-response.

Another example given by Schafer [153] is the median and 90th percentile for systolic blood pressure and total cholesterol (TCPMI). This code is shown in Figure 9.5. The first step is to define a blood pressure variable as an average of three measurements. The remainder of the code is very similar to the previous example, using with() to call svyby() to compute the quantiles.

For this analysis the results from R and SUDAAN are slightly different, due to different definitions of the median as described in section 2.4.1 and appendix C.4. In particular, there are some subgroups where the ties lead to the same estimated median for all five data sets and 0% estimated missing information. Discreteness of the data is noticeable here because the standard error of the quantiles is close to the resolution at which the data are recorded; systolic blood pressure is only measured to the nearest 2 mmHg and cholesterol to the nearest 1 mg/dl. The results are close to those from SUDAAN when the standard errors for quantiles are calculated using replicate weights.

```
> designs<-update(designs,
    systolic=(PEP6G1MI+PEP6H1MI+PEP6I1MI)/3)
> qres <- with(subset(designs, age>=20),
    svyby(~systolic+TCPMI, ~agegp+HSSEX, svyquantile,
        quantiles=c(0.5,0.9),se=TRUE))
> summary(MIcombine(qres),digits=2)
Multiple imputation results:
    with(subset(designs, age >= 20),
        svyby(~systolic+TCPMI, ~agegp+HSSEX, svyquantile,
            quantiles = c(0.5, 0.9), se = TRUE, design = .design))
    MIcombine.default(qres)
                             results  se (lower upper) missInfo
[20,40).1:0.5:systolic       117 0.38    116    117      0 %
[40,60).1:0.5:systolic       122 0.34    121    123      0 %
[60,Inf).1:0.5:systolic      134 0.88    133    136     22 %
[20,40).2:0.5:systolic       107 0.34    106    107      0 %
[40,60).2:0.5:systolic       117 0.60    116    118     38 %
[60,Inf).2:0.5:systolic      137 0.82    135    138     17 %
[20,40).1:0.5:TCPMI          187 1.39    184    191     85 %
[40,60).1:0.5:TCPMI          210 1.03    208    212     38 %
[60,Inf).1:0.5:TCPMI         210 1.20    207    212     18 %
[20,40).2:0.5:TCPMI          183 0.76    182    185     46 %
[40,60).2:0.5:TCPMI          208 1.50    205    211     41 %
[60,Inf).2:0.5:TCPMI         229 1.19    227    232     28 %
[20,40).1:0.9:systolic       132 1.49    129    135      7 %
[40,60).1:0.9:systolic       145 1.44    142    148      5 %
[60,Inf).1:0.9:systolic      163 1.38    161    166      4 %
[20,40).2:0.9:systolic       121 1.28    119    124      0 %
[40,60).2:0.9:systolic       141 1.34    139    144     16 %
[60,Inf).2:0.9:systolic      169 1.48    166    172     13 %
[20,40).1:0.9:TCPMI          243 1.86    239    247     13 %
[40,60).1:0.9:TCPMI          264 2.71    259    269      2 %
[60,Inf).1:0.9:TCPMI         265 3.38    258    272      0 %
[20,40).2:0.9:TCPMI          233 1.86    230    237      7 %
[40,60).2:0.9:TCPMI          271 2.32    267    276      1 %
[60,Inf).2:0.9:TCPMI         288 3.10    282    294     16 %
```

Figure 9.5 Quantiles of systolic blood pressure and total serum cholesterol in NHANES III imputation data

```
library(mitools)
library(RSQLite)
sqlite<-dbDriver("SQLite")
dbcon<-dbConnect(sqlite, dbname="~/nhanes/imp.db")

impload<-function(varlist,conn){
  tables<-paste("set",1:5,sep="")
  vars <- paste(varlist, collapse=",")
  query<-paste("select",vars,"from @tab@")
  data<-lapply(tables,
     function(table) dbGetQuery(conn, sub("@tab@",table,query)))
  imputationList(data)
 }
regdata<- impload(c("SDPPSU6","WTPFQX6","SDPSTRA6", "HSAGEU",
    "HSAGEIR","DMARETHN","BMPWTMI","BMPHTMI","DMPPIRMI",
    "HAB1MI","HAT28MI","HSSEX"), dbcon)
wtnames<-paste(paste("WTPQRP",1:52,sep=""),collapse=",")
regwts<- dbGetQuery(dbcon, paste("select",wtnames,"from core"))
designs <-svrepdesign(data=regdata, repweights=regwts,
    type="Fay", rho=0.3, weights=~WTPFQX6,combined=TRUE)
```

Figure 9.6   Loading data and replicate weights for a logistic regression model

The final example analysis is a logistic regression. The results presented by Schafer [153] use replicate weights. For this analysis we will load the necessary variables into memory explicitly. Code for loading the data is shown in Figure 9.6. After loading the necessary packages, the code sets up a connection to the database that holds the multiple imputations data. The function impload constructs a SQL select statement that reads a list of variables from each imputed data set in the database, and then creates a list of imputations from these.

In this example the replicate weights are identical in all five imputations and so a single copy of the weights is sufficient. These are loaded from the core database table and called regwts. If post-stratification or raking had been done after imputation the replicate weights might be different across imputations and they would then be loaded into an imputationList in the same way as the analysis variables.

The call to svrepdesign() constructs a list of five survey design objects with replicate weights. The replicate weights for the NHANES III imputations were constructed using Fay's method (Judkins [67]), a variant of BRR that multiplies the weights for each half-sample by $2 - \rho$ and ρ rather than 2 and 0, giving better stability for estimates in small categories.

Figure 9.7 shows code for constructing new variables and fitting a logistic model to predict being overweight. Age is coded into groups, body mass index is computed and then used to define overweight, and a poverty indicator variable is defined by dichotomizing the poverty income ratio at the poverty line. Self-rated health and

```
designs<-update(designs,
    age=ifelse(HSAGEU==1, HSAGEIR/12, HSAGEIR))
adults <- subset(designs, age>=20)
adults <-update(adults,
    race= factor(ifelse(DMARETHN==4, 1, DMARETHN)),
    bmi =BMPWTMI / (BMPHTMI/100)^2,
    poverty= DMPPIRMI<=1,
    evgfp = factor(HAB1MI),
    activity = factor(HAT28MI, levels=c(3,1,2,-9)),
    agegp = cut(age, c(0,20,40,60,Inf),right=FALSE)
    )
adults <- update(adults,
    overwt = ifelse(HSSEX==1, bmi >= 27.8, bmi >= 27.3))
models<-with(adults,
    svyglm(overwt~HSSEX+agegp+race+evgfp+activity+poverty,
        family=quasibinomial))
summary(MIcombine(models), digits=3)
```

Figure 9.7   Defining variables and fitting a logistic regression model

	results	se	(lower	upper)	missInfo
(Intercept)	-1.5902	0.1122	-1.8106	-1.370	10 %
HSSEX	0.0491	0.0507	-0.0504	0.149	5 %
agegp[40,60)	0.6964	0.0549	0.5885	0.804	9 %
agegp[60,Inf)	0.6065	0.0740	0.4614	0.752	4 %
race2	0.4481	0.0573	0.3358	0.560	2 %
race3	0.4103	0.0657	0.2815	0.539	3 %
evgfp2	0.4281	0.0702	0.2904	0.566	4 %
evgfp3	0.6852	0.0761	0.5359	0.834	6 %
evgfp4	0.7669	0.0825	0.6052	0.929	3 %
evgfp5	0.3757	0.1175	0.1453	0.606	6 %
activity1	-0.4014	0.0622	-0.5239	-0.279	14 %
activity2	0.2732	0.0564	0.1626	0.384	3 %
povertyTRUE	-0.0160	0.0637	-0.1421	0.110	18 %

Figure 9.8   Logistic regression predicting overweight from health and demographic variables

level of activity are turned into factors, with activity having 3 "About the same as others" as the reference level.

As before, with() is used to apply the same analysis to all five imputations, returning a list of results that are combined using MIcombine(). The results are shown in Figure 9.8.

EXERCISES

9.1   Fit regression models to examine how self-rated general health (HAB1MI) is related to age (HSAGEIR) and obesity (bmi as defined in Figure 9.7)

   a) using the NHANES III multiply imputed data, with replicate weights,

   b) using just the complete data before imputation (as described in subsection 9.2.2).

9.2   Repeat the logistic regression analysis in Figure 9.7 using linearization instead of replicate weights.

9.3   In the NHANES III imputation data, estimate mean, median, and 90th percentile of systolic blood pressure and total cholesterol broken down by rural/urban location and region of the country.

9.4   Fit regression models to examine how self-rated general health (HAB1MI) is related to obesity (bmi as defined in Figure 9.7)

   a) using just the complete data before imputation (as described in subsection 9.2.2),

   b) calibrating on age, sex, race, urban/rural location and region of the country (as described in subsection 9.2.2).

9.5   ⋆ For the incomplete data sets defined in Figure 9.1, fit a linear model to predict api00 from the other variables and impute from this regression model by taking the fitted value from the regression and adding a randomly selected residual.

   a) Impute a single complete data set and do the two analyses (mean and regression model) at the end of Figure 9.1. Compare the results to the true population values.

   b) Create 10 multiply-imputed data sets and repeat the two analyses. Compare the results to the true population values.

CHAPTER 10

★ CAUSAL INFERENCE

In which things might have turned out differently.

Another name for design-based inference is *randomization inference*, and the link to randomized experiments suggested by this term is real and useful. A randomized clinical trial, for example, selects one of two possible treatments for each individual and observes the outcome. It is possible, if slightly unnatural, to think of this procedure as selecting one of two potential outcomes for each individual: the outcome on treatment A and the outcome on treatment B. Based on the sample of one outcome for each individual, we wish to estimate the average difference between the treatment A outcome and the treatment B outcome. In the randomized trial the sampling of potential outcomes has a constant sampling probability of $\pi_i = 1/2$, so there is no need to take account of the sampling when performing inference.

In observational data we can still think of an exposure to A or B as being the result of a probability sampling procedure, just as in the previous chapter we thought of missing data in this way. For example, an individual i has an unknown probability π_i of being a smoker and $1 - \pi_i$ of being a non-smoker. We can pretend that the smoking status we see resulted from a randomization with these probabilities, and that

Complex Surveys: A Guide to Analysis Using R. By Thomas Lumley
Copyright © 2010 John Wiley & Sons, Inc.

the outcomes we see are a probability sample of the potential outcomes according to these probabilities. The actual effect of smoking is the population average difference between the "smoker" and "non-smoker" outcomes for an individual, and we can estimate this by the techniques developed in earlier chapters. The use of potential outcomes in this way can either be taken as part of a definition of cause and effect or as a useful tool in constructing estimators.

From experience with complex designs it is clear that the sampling probabilities matter. Ignoring the sampling probabilities and simply taking the difference in unweighted mean outcome between smokers in the sample and non-smokers in the sample could be very misleading about the population effect. Comparing a potential-outcomes view of cause and effect to a missing data problem makes it clear how difficult causal inference really is: it is like estimation with 50% missing data. The analogue of the necessary but implausible "Missing At Random" assumption is the assumption of "no unmeasured confounding"; that the analysis measures and includes all variables that affect both exposure and outcome. It will never be possible to draw definitive conclusions about cause and effect solely from observational data, but a lot can be learned with care. The main reason why researchers spend lifetimes collecting observational data, and taxpayers fund them to do so, is because knowledge about cause and effect is gained that way.

This chapter will present only a brief introduction to inverse-probability of treatment weighting and causal inference. The sampling approach that we discuss is only one of the mathematical approaches to the problem, and the characterization of causality in terms of potential outcomes is likewise not the only option. The explicit modelling of potential outcomes in statistics was introduced by Rubin [140, 142] and is examined in detail by Holland[62] and his discussants. Pearl [124] describes a translation between causal claims and statements about graphical networks and its implications for inference. A discussion of some of the philosophical issues is given by Cartwright [30]. Further statistical development and related estimators can be found in Frangakis & Rubin [48] and in the papers of Robins and co-workers — perhaps most accessibly in those with Hernán.

10.1 IPTW ESTIMATORS

The Inverse Probability of Treatment Weighted (IPTW) estimator is the analog for potential-outcomes inference of the Horvitz–Thompson estimator. We first examine how IPTW estimators of a treatment effect work in randomized trials, with known sampling probabilities. We then consider how to extend these methods to observational data by analogy with methods for non-response adjustment in probability samples.

10.1.1 Randomized trials and calibration

In a randomized trial the treatment probabilities for each individual are known. For simplicity we consider the typical case of equal randomization to two arms, a treatment or a control, so that $\pi_i = 1/2$ for each individual. The IPTW estimator of the treatment effect on the mean of an outcome variable Y is the difference in the estimated means between treatment and control groups. Since the sampling weights are the same for everyone, the IPTW estimate of the mean is the same as the unweighted mean, and the IPTW estimate of the treatment effect is just the difference in the mean of Y between treatment and control groups.

If a variable X, measured before randomization, is a strong predictor of Y it is standard practice to estimate the treatment effect by fitting a linear model with X and treatment as predictors, often called an *ANCOVA* model. Another way to use X is a calibration estimate with X, treatment, and $X \times$ treatment as auxilliary variables.

It is convenient to code treatment as $-1/2$ for the treatment group and $+1/2$ for the control group (rather than 0 and 1). With this coding, the total of the treatment variable over all potential outcomes is exactly zero; each person has both a treatment ($X = +1/2$) outcome and a control ($X = -1/2$) potential outcome. The total for $X \times$ treatment is also zero — since X is a pre-randomization variable its value is the same for the treatment potential outcome and the control potential outcome for each individual, and after multiplication by the treatment variable these will cancel. Finally, the total for X over all $2n$ potential outcomes is twice the sum of X over the n observed outcomes, since X is the same for both potential outcomes in each individual.

It should not be surprising given the relationship between calibration and linear regression that the ANCOVA estimate and the calibrated IPTW estimate are closely related. In fact they are identical, as shown in the Appendix (section A.6). This provides another useful analogy for explaining to biostatisticians how and why calibration increases precision.

When the treatment effect is summarized in some way other than a difference in means, such as a relative risk, odds ratio, or hazard ratio, the calibration estimate will typically not be identical to a regression estimate adjusting for pre-randomization variables. For the odds ratio and hazard ratio, in particular, the adjusted estimate from a regression model does not estimate the same parameter as an unadjusted estimate; it is further away from the null value of 1.0, a phenomenon called *non-collapsibility* [53]. The calibration estimate does estimate the same parameter as the unadjusted estimate, but with higher precision. The overall impact on power of adjustment and calibration seems to be similar.

Figure 10.1 shows code that simulates the adjustment and calibration estimators of an odds ratio for a binary outcome. The code first simulates the data based on treatment (`trt`) and a baseline variable (`x`) and then constructs a survey design object. The auxiliary variable for calibration (`z`) is a transformation of `x` that in practice would have to be estimated. The sample size for the trial is 2000, with an average of 270 individuals having the event $Y = 1$.

The true adjusted log odds ratio is 0.5 as provided in the simulation code. To compute the true unadjusted log odds ratio we need to compute the probability of $Y = 1$ in the treatment group and in the control group, by averaging over the

```
trt<-rep(c(-1/2,1/2),1000)
one.sim<-function(){
  x <- rnorm(2000)
  mu <- expit(trt/2+x*2-3)
  y <- rbinom(2000, 1, mu)
  df <- data.frame(y=y, x=x, trt=trt, weight=2, z=expit(x*2-3))

  des <- svydesign(id=~1, weight=~weight, data=df)
  cdes <- calibrate(des, ~z*trt, pop=c(4000,2*sum(df$z),0,0))

  m0 <- svyglm(y~trt, family=binomial, design=des)
  m1 <- svyglm(y~trt+x, family=binomial, design=des)
  m2 <- svyglm(y~trt, family=quasibinomial, design=cdes)

  sapply(list(m0, m1, m2),
    function(m)c(beta=coef(m)["trt"], se=SE(m)["trt"]))
}
```

Figure 10.1   Calibration and adjustment for baseline in logistic regression models for a simulated randomized trial

distribution of x:

$$\Pr[Y = 1|\text{treatment}] = E\left[\Pr[Y = 1|X = x, \text{treatment}]\right]$$
$$\Pr[Y = 1|\text{control}] = E\left[\Pr[Y = 1|X = x, \text{control}]\right]$$

The averages can be computed as 15.0% in the treatment group and 11.1% with the `integrate()` function giving the true unadjusted log odds ratio as 0.345.

Table 10.1 shows the results of 1000 simulations. The unadjusted and calibrated parameter estimate have mean value very close to the true log odds ratio of 0.345, and the adjusted parameter estimate matches the true adjusted odds ratio of 0.5. The mean of the estimated standard errors is matches the actual standard error of the simulated results, showing that the estimated standard errors are reliable. The power of the test for treatment effect increases from 74% for the unadjusted estimate to 88% for both the adjusted and calibrated estimates. The calibration approach has the advantage of estimating the same quantity that is estimated by the unadjusted estimate, which simplifies the interpretation of results and allows easier comparison between studies.

The same calibration estimators have been derived by Tsiatis and coworkers [174, 192] based on semiparametric theory rather than on sampling from potential outcomes. As in our development they were motivated by estimating the same parameter more precisely, avoiding the change in the target of inference that occurs when adjusting for covariates in non-linear models. They also showed that the choice of auxiliary variables can be carried out separately for each treatment group, by

Table 10.1   Results of calibration and adjustment for baseline in logistic regression models for a simulated randomized trial (based on 1000 simulations)

	mean $\hat{\beta}$	mean $\widehat{SE}$	simulation SE	Power
Unadjusted	0.343	0.134	0.132	74%
Adjusted	0.500	0.162	0.161	88%
Calibrated	0.345	0.112	0.111	88%

separate analysts blinded to treatment, so that the analysts could not tell whether a particular choice of auxiliary variables would increase or decrease the estimated treatment effect.

The calibration approach to randomized trials is also similar to an existing approach called 'non-parametric analysis of covariance' (Tangen and Koch [167], Koch et al. [78]), but calibration may be easier to understand. Tangen and Koch replace the logistic regression estimation procedure by a population mean of influence functions, as we did in constructing auxiliary variables for regression models in section 8.5.1, and then use a regression estimator of the population mean of the influence functions

10.1.2   Estimated weights for IPTW

In observational data we do not know the sampling probabilities π_i for the potential outcomes; they must be estimated from the observed data just as we did for non-response in the previous chapter.

As an example we use a set of measurements of lung function on 654 children, from 290 families, in the file kidsmoking.txt. These data were collected by Tager et al. [166] in Boston, in the 1970s and used as an example by Rosner[139]. A sample of families was taken by sampling children aged 5–9 from schools in East Boston, and then recruiting the families of sampled children. In this analysis we will treat the data as being a simple random sample of families, although families with more children aged 5–9 would have been more likely to be recruited.

The lung function measurement is FEV_1, forced expiratory volume in 1 second. This is the amount of air the child can blow out in one second, and it is a useful measure both of lung size and of breathing problems such as asthma. The data set also includes height, age, sex, and whether the child smokes cigarettes. We are interested in the impact of smoking on FEV_1. In adults, smoking is known to reduce FEV_1, but there is less data in children.

Figure 10.2 shows a boxplot of FEV_1 and smoking. There is a large difference between smokers and non-smokers, and a t-test (with the t.test() function) gives a p-value of 3×10^{-10}. It is tempting to jump to the causal conclusion that this difference is an effect of smoking, at least until one checks the coding of the smoking variable and finds that the smokers have *higher* FEV_1 than the non-smokers. The difference is almost entirely due to differences in age, as older children, with larger

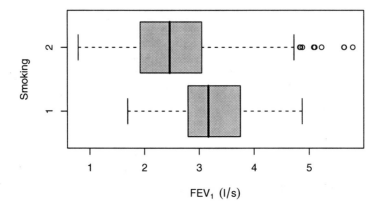

Figure 10.2   Forced Expiratory Volume (1sec) and smoking status for 654 children.

lungs, are also more likely to be smokers. Figure 10.3 shows a scatterplot of FEV_1 and age. The difference between smokers and non-smokers at the same age is much less dramatic, but there is a suggestion of lower FEV_1 for smokers at the older ages — the lowest values are in smokers and the highest in non-smokers.

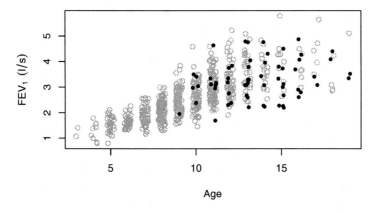

Figure 10.3   Forced Expiratory Volume (1sec) and age. Solid circles indicate smokers. Age has been jittered to reduce overlap

The first step in an IPTW analysis is to estimate the probability of being a smoker, as a function of the available data. Figure 10.4 shows the R code for this estimation. As there are no smokers younger than nine years and only one younger than 10 there cannot be any real loss of information in restricting to children 10 and older. There are two simple models (models m1 and m2) that predict FEV_1 about equally well, an age by sex interaction or an age by height interaction. Since the children were sampled by family it is also useful to consider predictors of smoking based on other children in the family. We can define a variable sib that indicates whether the child has a

```
kids <- read.table("kids.txt", header=TRUE)
olderkids <- subset(kids, age>9)
olderkids$is_smoker <- 2-olderkids$smoke
olderkids$male <- 2-olderkids$sex

m1 <- glm(is_smoker~age*male, data=olderkids,
   family=binomial)
m2 <- glm(is_smoker~height*male, data=olderkids,
   family=binomial)
m1a <- glm(is_smoker~(age+sib)*male, data=olderkids,
   family=binomial)
m2a <- glm(is_smoker~(height+sib)*male, data=olderkids,
   family=binomial)

olderkids$pi1 <- ifelse(olderkids$is_smoker,
     fitted(m1), 1-fitted(m1))
olderkids$pi1a <- ifelse(olderkids$is_smoker,
     fitted(m1a), 1-fitted(m1a))
olderkids$pi2 <- ifelse(olderkids$is_smoker,
     fitted(m2), 1-fitted(m2))
olderkids$pi2a <- ifelse(olderkids$is_smoker,
     fitted(m2a), 1-fitted(m2a))
d1 <- svydesign(id=~family, prob=~pi1, data=olderkids)
d2 <- svydesign(id=~family, prob=~pi2, data=olderkids)
d1a <- svydesign(id=~family, prob=~pi1a, data=olderkids)
d2a <- svydesign(id=~family, prob=~pi2a, data=olderkids)

summary(svyglm(fev1~is_smoker, design=d1))
summary(svyglm(fev1~is_smoker, design=d1a))
summary(svyglm(fev1~is_smoker, design=d2))
summary(svyglm(fev1~is_smoker, design=d2a))
```

Figure 10.4   IPTW estimation for the effect of smoking on lung function in children

Table 10.2 Estimated effects of smoking on FEV_1 in children, from IPTW analysis

	unweighted	m1	m1a	m2	m2a
Age 10 and older	0.15	0.11	0.19	0.03	0.17
(standard error)	(0.13)	(0.13)	(0.15)	(0.16)	(0.13)
Age 12 and older	-0.18	-0.13	-0.14	-0.17	-0.11
(standard error)	(0.16)	(0.16)	(0.18)	(0.16)	(0.16)

sibling who is a smoker, and this improves prediction in both the height-based and age-based models, giving models m1a and m2a. The fitted probabilities of smoking in these models (using the function fitted()) range from about 1% up to about 75%, suggesting that a simple comparison of means even in this restricted age range may well be unreliable.

The fitted probabilities p_i of smoking are not the same as the sampling probabilities π_i that we need for weighting. The sampling probability is the probability of sampling the potential outcome that we actually observed, rather than the other potential outcome. For smokers, we observe the smoking outcome, so the sampling probability is the probability of smoking, $\pi_i = p_i$. For non-smokers we observe the non-smoking outcome, so the sampling probability is the probability of not smoking, $\pi_i = 1 - p_i$. The sampling probabilities, in the variables pi1, pi1a, pi2, and pi2a, range from about 0.05 to 0.99. Four survey design objects are defined, corresponding to the four models. These incorporate the cluster sampling by family as well as the estimated sampling weights for potential outcomes. Finally, the calls to svyglm() estimate the effects of smoking. As the estimated sampling weights were specified directly to svydesign() rather than derived by calibration the standard error estimates for IPTW estimates will be conservative, not reflecting any precision gains from calibration.

Table 10.2 shows the estimated effects of smoking for these four models, and also the estimated effects restricted to children 12 and older. When all children aged 10 and older are analyzed there is still a suggestion of higher FEV_1 in smokers. This is likely to be a confounding effect by a combination of height and age, as it disappears when the analysis is restricted to age 12 and older. There is also less variation between models in the estimated effect in this older subset. If we believe the analysis in children aged 12 and older, there is weak evidence for a small reduction in FEV_1 from smoking. The estimates are very similar to those from an unweighted analysis of all children in the specified age ranges. This suggests that the large variation in smoking probability is largely independent of FEV_1 in these children. Figure 10.3 shows that age is weakly related to FEV_1 after about age 12, although strongly related to smoking. Similarly, whether a sibling smokes is strongly related to smoking status but more-or-less independent of FEV_1. The IPTW analysis is still valuable; without doing it we would have no basis for assuming that the unweighted analysis was approximately correct.

Table 10.3 Estimated effects of smoking on FEV_1 in children, from double-robust analysis

	unweighted	m1	m1a	m2	m2a
Age 10 and older	−0.06	−0.15	−0.10	−0.04	0.05
standard error	(0.09)	(0.10)	(0.11)	(0.10)	(0.10)

10.1.3 Double robustness

A more typical approach to estimating the effect of smoking in children would be to fit a linear regression model to FEV_1 and adjust for height, sex, and other relevant variables. These two approaches can often be combined, just as they were for non-response in Chapter 9. The result is a doubly-robust estimator of the effect of exposure: it will be valid if *either* the regression model *or* the IPTW reweighting is successful in removing confounding[136] .

To illustrate the double robust estimator, we use the same sets of sampling weights for children aged 10 and over as in Table 10.2 but now fit models for effect of smoking adjusted for height and sex. The results are in Table 10.3. The apparent beneficial effect of smoking that we saw in this age group in Table 10.2 has gone.

As was the case for non-response, a double-robust analysis is not always possible for technical reasons. For example, there is no double-robust version of logistic regression for a binary outcome. It is necessary that the regression model for the outcome and the regression model for the exposure probabilities are mathematically compatible; that they could both be true simultaneously. It is still reasonable to model both the outcome and the exposure probability and to hope for an increase in model robustness by doing so. Robins and Rotnitzky [136] use the term *generalized doubly-robust* for estimators constructed from technically incompatible potential-outcome and regression models.

10.2 MARGINAL STRUCTURAL MODELS

For estimating effects of a single binary exposure there is not much to choose between IPTW estimators and regression estimators. Both will be valid if there are no unmeasured confounders, and any measured confounder can equally well be used in adjustment, in reweighting, or in both. With non-linear models such as logistic regression it may be convenient that the IPTW estimator targets the same parameter as the model changes, where a regression estimator targets different parameters (as in section 10.1.1), but this is a relatively minor point.

With repeated exposures over time there is a real advantage in IPTW-type estimators, in handling feedback from outcome to exposure. In a longitudinal study with exposure and outcome measured at times 1, 2, 3, 4, for example, outcome at time 2 may affect exposure at time 3 and so be a confounder. A regression model for the overall effect of treatment would have to adjust for outcome at time 2. On the other hand, outcome at time 2 is affected by exposure at time 1 and so is part of the pathway by which treatment acts. Adjusting for these intermediate outcomes changes the target of estimation so that regression will no longer estimate the true effect of treatment. The fact that IPTW reweighting removes confounding *without changing the target of estimation* is now crucial, since it makes it possible to bypass this problem. The resulting *marginal structural models* allow the effect of any fixed sequences of exposures to be compared.

For example, consider a study comparing the effects of different blood pressure drugs on cognitive function in elderly patients. In addition to any direct effects of the drugs on the brain there are effects of blood pressure, since high blood pressure can cause strokes or damage the small blood vessels in the brain. If a particular drug is unsuccessful in controlling blood pressure the treatment will be changed, so that blood pressure at intermediate time points is both a confounder for future treatment and part of the effect mechanism for earlier treatments. Adjusting for blood pressure in a regression model would remove confounding but will also remove the indirect effects from the target of estimation. Reweighting and using an IPTW estimator will remove confounding without masking the blood-pressure effects on cognitive function. More examples and references, and a readable overview of these models can be found in two papers by Robins, Hernán, and Brumback, [135, 60] and a worked example is presented by Cook et al. [36]

With a binary exposure at four time points there are $2^4 = 16$ possible sequences of exposure and so 16 potential outcomes for each person, only one of which is observed. As we did with a single exposure in section 10.1.2, the first step is to construct a model for the probability of each exposure sequence. A straightforward approach is to fit separate logistic models for the probability of exposure at each time point, using past exposure, outcomes, and other covariates as predictors. The fitted probabilities at each time point can then be multiplied together to give probabilities for any exposure sequence. The sampling probability for the observed data is just the probability for the exposure sequence that was actually observed.

Example: maternal stress and children's illness. The Mothers' and Children's Morbidity Study examined the relation between maternal stress and childhood illness. Daily observations of stress and illness were made for 30 days on 167 mother–child pairs. These data were analyzed by Zeger and Liang [191] using logistic regression, despite the possibility that childhood illness could also affect maternal stress, giving rise to feedback confounding. The data were reanalyzed by Robins et al. [134] using a causal inference technique called the "G–computation algorithm." We will analyse the data with a marginal structural model, *i.e.*, with logistic regression models reweighted using IPTW.

We assume that the mother's stress on a particular day does not affect the child's health on that day, and we want to evaluate whether it affects future illness. The assumption of no same-day effect is necessary for the analysis. To the extent that illness is evaluated objectively, the assumption is plausible on biological grounds, but maternal stress could potentially affect the perception of illness in the child.

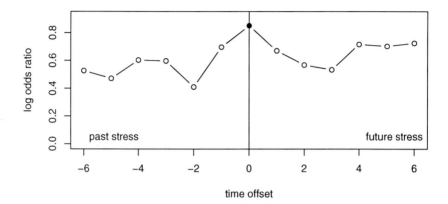

Figure 10.5 Log odds ratio between childhood illness and mother's past, current, and future stress

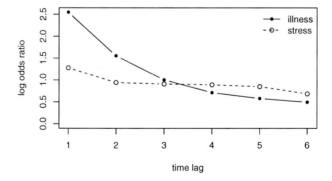

Figure 10.6 Log odds ratios for association at increasing time lags, for mother's stress and child's illness

```
mscm <- read.table("mscm.txt")
names(mscm) <- c("id","day","stress", paste("s",1:6,sep=""),
    "illness",paste("i",1:6,sep=""),
    "married","education","employed","chlth","mhlth","race",
    "csex","hsize","wk1illness","wk1stress")
mscm$nna <- with(mscm, ave(stress+illness, id,
  FUN=function(x) sum(!is.na(x))))
mscm <- subset(mscm, nna==28 & day>6 & day<29)

model <- glm(stress~illness+i1+s1+day+wk1stress, family=binomial,
    data=mscm,na.action=na.exclude)
mscm$pstress <- fitted(model)
basemodel <- glm(stress~day+wk1stress, family=binomial,
    data=mscm, na.action=na.exclude)
mscm$pbstress <- fitted(basemodel)

mscm$swit <- with(mscm, ifelse(stress==1,
      pbstress/pstress,(1-pbstress)/(1-pstress)))
mscm$swi<-with(mscm,ave(swit, id, FUN=prod))

des<-svydesign(id=~id, weights=~swi, data=mscm)
unwt<-svydesign(id=~id, data=mscm)
```

Figure 10.7   Estimating probability of exposure to set up a marginal structural model for child's illness and mother's stress

Figure 10.5 summarizes the raw association between mother's stress and child's health by the log odds ratio for the 2×2 table of illness by stress on the current day, six days into the past, and six days into the future. The odds ratio is highest on the same day, indicating some relationship between the two series, and then falls off. It appears to have stabilised by about four days lag. Figure 10.6 shows the assocation over time for each variable separately, again with log odds ratios. Mother's stress shows a relatively flat pattern, indicating that some mothers are consistently more stressed than others. Illness shows a steep decline, indicating that illness happens in relatively short periods that are then over with. Based on these plots it looks as though six days of history is sufficient for modelling.

Figure 10.7 shows code for setting up the data, fitting the logistic regression models for exposure, and computing the weights. After reading the data, the first step is to remove observations with too many missing values, and also to remove observations from the first week where the history is not available.

The logistic regression model model predicts maternal stress using illness on the current and previous day, time, and average stress during the first week. The second logistic regression, to fit basemodel uses only time and average stress during the first week, that is, it uses only deterministic or baseline variables. The reason for

```
> m_iptw<-svyglm(illness~s1+s2+s3+s4, des=des,
    family=quasibinomial)
> m_raw<-svyglm(illness~s1+s2+s3+s4, des=unwt,
    family=quasibinomial)
> m_gee<-svyglm(illness~s1+s2+s3+s4+wk1stress+wk1illness,
    des=unwt, family=quasibinomial)
>
> pred_data <- data.frame(s1=0:1, s2=0:1, s3=0:1, s4=0:1,
    wk1stress=0.16, wk1illness=0.16)
>
> pred_iptw<-predict(m_iptw, pred_data,
    vcov=TRUE, type="response")
> pred_raw<-predict(m_raw, pred_data,
    vcov=TRUE,type="response")
> pred_gee<-predict(m_gee, pred_data,
    vcov=TRUE,type="response")
>
> svycontrast(pred_iptw,c(-1,1))
         contrast     SE
contrast 0.094968 0.1208
> svycontrast(pred_raw,c(-1,1))
         contrast     SE
contrast  0.21593 0.0943
> svycontrast(pred_gee,c(-1,1))
         contrast     SE
contrast  0.15859 0.0862
```

Figure 10.8   Estimating the effect of stress on illness in data from Mothers' and Children's Morbidity Study

this model is to stabilize the weights: weighting by any deterministic or baseline variables cannot affect feedback confounding and will increase the variability of the weights. Dividing the weights from the full model for maternal stress by the weights derived from this baseline model will reduce the variability of the weights while still preserving the lack of confounding.

The variable `swit` is the contribution to the stabilized weight from a single observation. It is the sampling probability estimated from `basemodel` divided by the sampling probability estimated from `model`. These contributions are then multiplied together within each individual to give the per-individual stabilized weight in the variable `swi`. Finally, we define two survey design objects. The first, `des`, uses the stabilized weights, the second, `unwt`, is unweighted. Both specify that the data are cluster-sampled on `id` so that standard errors will correctly incorporate within-individual correlations.

Figure 10.8 shows the results of three analyses of the data. The first is the IPTW analysis using the stabilized weights. The second fits the same model to the unweighted data without any adjustment for confounding. The third attempts to remove for confounding by adjusting for the baseline illness and stress measures. Rather than interpreting the coefficients of these models, we construct the predicted probabilities of illness that the models imply, comparing a child whose mother has been stressed for four days to one whose mother has been unstressed for four days, and using the sample average value for the baseline covariates.

The effect estimate from the unadjusted, unweighted model is 0.216, that is, the probability of illness is 21.6 percentage points higher in the child whose mother was stressed. Adjusting for the baseline covariates brings the difference down to 15.8 percentage points, but this is still confounded by the time-dependent effects of illness on stress. The IPTW analysis gives an estimate of 9.5 percentage points. Robins et al. [134] estimated 10.2 percentage points using a different but related method to handle feedback confounding.

The estimate of 9.5 percentage points difference is not statistically different from zero. Part of the problem is a small sample size, and this is aggravated by the large difference in weights. Truncating the largest weights to reduce the influence of a few points is often a useful strategy, although there is a risk of reintroducing confounding. In this example it does not make any real difference. Truncating the range of the weights to 50 gives an estimate of 6.8 percentage points and a standard error reduced from 12 to 10 percentage points.

Marginal structural models often give relatively disappointing results as in this example, because causal inference from observational data is very hard and uncertainties are genuinely large. The most promising applications are situations with large samples and where a single intermediate variable has a large effect on future exposure and is very well measured. The papers by Robins, Hernán, and Brumback [60, 135] analyze data on HIV treatment from the Multicenter AIDS Cohort Study in the early years of the epidemic, where AZT was the only available treatment and where it was used in response to decline in CD4 counts. The data set is large, CD4 counts are well measured, and they were the primary criterion used in starting treatment.

Appendix A

ANALYTIC DETAILS

A.1 ASYMPTOTICS

A.1.1 Embedding in an infinite sequence

For an estimate $\hat{T}$ of a population quantity T to be useful we need to know that $\hat{T}$ is close to T with high probability. For most purposes we also need to know how close they are. It is rarely feasible to study the exact distribution of survey estimators analytically except in trivial cases, so some form of approximation is needed. Most commonly, we try to demonstrate that the distribution of $\hat{T} - T$ under repeated sampling from the same population is close to $N(0, \hat{\sigma}^2)$, where $\hat{\sigma}$ is the estimated standard error. This approximation has two parts, first that the distribution of $\hat{T} - T$ is close to Normal, and second that $\hat{\sigma}^2$ is close to the variance of the distribution.

Asymptotic approximations work by discarding differences that will be small in large samples. Formally, the actual population and sampling design are embedded in an infinite sequence of populations and sampling designs and the approximations are

*

Complex Surveys: A Guide to Analysis Using R. By Thomas Lumley
Copyright © 2010 John Wiley & Sons, Inc.   **217**

the limits taken along the sequence. Rather than analyzing a particular population value T and estimate $\hat{T}$ we work with a sequence of true population values T_n and estimates $\hat{T}_n$. For this to be useful the populations in the infinite sequence have to be sufficiently similar to the actual population, so that T_n can meaningfully be compared for different populations in the sequence. We care about the infinite sequence only to the extent that it is informative about what happens in real populations.

One way to construct infinite sequences of populations is to regard the Nth population in the infinite sequence as an independent and identically distributed sample of size N from a fixed superpopulation model. The advantage of this approach is that most population summary statistics will converge to a constant (e.g., the mean or median) or to a constant multiple of some known function of N (e.g., the total, which is proportional to N). The disadvantage is that the resulting sequence is too well-behaved, and may not provide good asymptotic approximations for situations such as a sample with many small strata. This approach does make sense, however, for two-phase sampling where the first phase is modelled as independent and identically distributed sampling from a model for some process (a superpopulation model) and the second phase is probability sampling from the first-phase sample (e.g., Breslow, McNeney, Wellner[18])

Isaki and Fuller[65] constructed a nested infinite sequence of populations by taking an infinite sequence u_n of elements and defining the nth population to contain elements $\{u_1, \ldots, u_n\}$. The nth sample is taken from the n population independent of any other sample, so the samples are not nested. The nesting of populations ensures that the sequence of true population values will converge to a limit under fairly weak assumptions about the actual values involved and the sampling designs. Hájek [56] also used nested populations, with $u_i = i$, and did not impose any relationship between the samples.

A.1.2   Asymptotic unbiasedness

Let r_n be a sequence of normalizing constants such that

$$r_n(\hat{T}_n - T_n) = O_p(1)$$

and

$$r_n(\hat{T}_n - T_n) \neq o_p(1).$$

For example, when each population is a sample from the same superpopulation model and n/N converges to a constant, $r_n = \sqrt{n}/N$ for the estimated total and $r_n = \sqrt{n}$ for the mean. It is convenient to be able to ignore the bias in $\hat{T}_n$ in inference, so we want to define an *approximately unbiased* statistic as one where the bias can be ignored in large samples. Heuristically, we want something that says the bias is small compared to the standard error, along the lines of

$$E[\hat{T}_n - T_n] = o(r_n^{-1}),$$

but this cannot be used directly as a definition since $E[\hat{T}_n]$ is too sensitive to outliers, and may not even exist ($\hat{T}_n$ may be infinite with non-zero probability).

For example, consider the ratio estimator

$$\hat{R}_n = \frac{\hat{T}_Y}{\hat{T}_X} = \frac{\sum_{i=1}^n Y_i/\pi_i}{\sum_{i=1}^n X_i/\pi_i}.$$

If there is a central limit theorem for (X, Y) jointly, then for some sequence r_n

$$r_n(\hat{T}_{X,n} - T_{X,n}) \xrightarrow{d} N(0, \sigma_x^2)$$

and similarly for $\hat{T}_Y$.

If it is possible to have $X_i = 0$ and $Y_i > 0$ there is a positive probability that $\hat{T}_{Y,n} = 0$ and the bias of $\hat{R}_n$ is infinite. Even if X_i is bounded away from zero the ratio estimator is biased in finite samples. It is asymptotically unbiased in the sense that

$$\sqrt{n}(\hat{R}_n - R) \xrightarrow{d} N(0, \tau^2)$$

for some $\tau^2 > 0$, so that the mean of the limiting distribution is zero and for moderately large n the bias can be ignored. For example, the confidence interval construction $(\hat{R}_n - 1.96 \times \hat{\tau}, \hat{R}_n + 1.96 \times \hat{\tau})$ will have close to 95% coverage, and the coverage error will go to zero as $n \to \infty$.

To capture this sort of asymptotic behavior we can define a (scalar) sequence of estimators $\hat{T}_n$ of population constants T_n as *approximately unbiased* when

$$r_n(\hat{T}_n - T_n) \xrightarrow{d} F \tag{A.1}$$

where F has zero mean and non-zero variance (and in practice will be Normal). This definition is based directly on the reason that we care about approximate unbiasedness: inference about T_n can be based on $\hat{T}_n$ and its standard error.

An alternative definition that does not explicitly use convergence in distribution works by truncating rare outliers. Define $\hat{T}_n^{(M)}$ as $\max(\min(\hat{T}_n, M), -M)$, and say that $\hat{T}_n$ is approximately unbiased if

$$\lim_{M \to \infty} \lim_{n \to \infty} r_n E\left[\hat{T}_n^{(M)} - T_n\right] = 0. \tag{A.2}$$

That is, the bias is made well-defined by truncating any rare, extreme outliers, and we require the bias to go to zero as the truncation limit increases.

These are almost equivalent: equation A.1 implies equation A.2 when F has enough finite moments, and equation A.2 implies that every subsequence of $\hat{T}_n$ has a further subsubsequence along which equation A.1 holds.

It is also important to note that there is another quite different property for which the name *asymptotic unbiasedness* is used (eg Chapter 5.3 of Särndal et al. [151]). Their definition is $E[\hat{T}_n - T_n] \to 0$. Ignoring for simplicity any issues about existence of expected values or interchange of limits, this property is much weaker than our definition, at least for statistics such as means, medians, and regression coefficients where $r_n \to \infty$. It does not imply consistency, nor does it imply (even together with assuming consistency and asymptotic Normality) that the bias can asymptotically be neglected. As far as I can tell, all the claims about asymptotic unbiasedness made in Särndal et al. [151] are still true for our stronger definition if in addition it is assumed that the estimate is asymptotically Normal. Särndal [149], following Kott[84], uses "nearly design unbiased" to refer to essentially the same property as equations A.1 and A.2.

A practical scenario where the definitions disagree is smoothing. An optimal estimate $\hat{f}(x)$ for a smooth curve $f(x)$ will be consistent and asymptotically unbiased in the weaker sense, but its asymptotic distribution will be given by

$$n^{2/5}(\hat{f}(x) - f(x)) \xrightarrow{d} N(\eta(x), \sigma^2(x))$$

where $\eta(x)$ and $\sigma(x)$ are of about the same size. The bias in the asymptotic distribution cannot be neglected and the estimator is not asymptotically unbiased in our stronger sense.

A.1.3 Asymptotic normality and consistency

If an asymptotic sequence is defined so that means are asymptotically Normal the continuous mapping theorem gives asymptotic normality and consistency for explicit functions of means. Standard "classical smoothness" arguments [e.g., Chapter 5 of van der Vaart [179]] give asymptotic normality and consistency for implicit functions that solve equations of the form

$$\sum_{i=1}^{n} \frac{1}{\pi_i} U_i(\theta) = 0$$

or maximize functions of the form

$$\sum_{i=1}^{n} \frac{1}{\pi_i} m_i(\theta)$$

where U_i or m_i depends only on the data from observation i. This includes linear and generalized linear models and loglinear models.

The Cox model is more complicated, because the estimating functions do not depend only on a single observation. The best-known asymptotic arguments for the Cox model rely on martingale theory and are not applicable to complex designs. These arguments rely heavily on the assumption that the model is correctly specified, and even with that assumption would not be valid for many complex designs such as cluster samples, or samples dependent on the outcome.

There still is no rigorous asymptotic theory for the Cox model under any reasonably general sampling design. The same limitations would apply to other semiparametric models whose mathematical theory relies on empirical process results. The best current results appear to be those of Breslow and Wellner [19, 20].

A.2 VARIANCES BY LINEARIZATION

The Horvitz–Thompson estimator gives an estimate of the variance of a total. If a statistic is explicitly a continuously differentiable function of one or more totals, the delta method gives an estimate of the asymptotic variance of the statistic. We have

$$\widehat{\text{var}}\left[\hat{f}(\hat{T}_1, \hat{T}_2, \ldots, \hat{T}_p)\right] \approx \sum_{i,j=1}^{p} \frac{\partial f}{\partial T_i} \widehat{\text{cov}}\left[\hat{T}_i, \hat{T}_j\right] \frac{\partial f}{\partial T_j}$$

where the relative error in the approximation goes to zero when the coefficients of variation of the totals go to zero.

Many statistics are not explicit functions of totals but are estimated by solving a system of equations that are estimated population totals. If $U_i(\theta)$ is a function of the data for observation i and a parameter θ, and the true population statistic θ^* solves the population equation

$$\sum_{i=1}^{N} U_i(\theta^*) = 0$$

then in a complex sample we would define $\hat{\theta}$ as solving the weighted sample equation

$$\sum_{i=1}^{n} \check{U}_i(\hat{\theta}) = 0.$$

If $\sum_i U_i(\theta)$ is sufficiently smooth the equation implicitly defines $\hat{\theta}$ as a function of $\sum_i U_i(\theta)$, with derivative matrix the inverse of the derivative matrix for U as a function of θ. Applying the delta method to this implicit function:

$$\widehat{\text{var}}\left[\hat{\theta}\right] \approx \left(\sum_{i=1}^{n} \frac{\partial U_i(\hat{\theta})}{\partial \theta}\right)^{-1} \widehat{\text{cov}}\left[\sum_{i=1}^{n} U_i(\hat{\theta})\right]\left(\sum_{i=1}^{n} \frac{\partial U_i(\hat{\theta})}{\partial \theta}\right)^{-1}. \tag{A.3}$$

This approach is called (Taylor Series) linearization, and is described by Binder[8]. It is closely analogous to the "classical smoothness" argument for estimators from estimating equations in model-based statistics.

A linearization argument for Cox model is more complicated, because the estimating functions U_i do not depend only on observation i. The variance calcuation is given by Binder [9], but rigorous proofs of consistency and asymptotic normality are still only available for restricted classes of designs (e.g., Lin[96], Breslow and Wellner[20])

A.2.1 Subpopulation inference

Variance estimation for a subpopulation total is most easily analyzed by defining a variable that is zero for individuals outside the subpopulation. If the goal is to estimate the population total of X over the subpopulation with $Z > 0$, then we define a variable $Y_i = X_i(Z_i > 0)$, and observe that the subpopulation total of X is the population total of Y. Using replicate weights, the variance of $\hat{T}_Y$ is computed from replicates $\hat{T}_Y^*$ and these are identical to replicates in which observations outside the subpopulation are simply dropped from the data:

$$\sum_{i=1}^{n} Y_i w_i = \sum_{Z_i > 0} X_i w_i.$$

The situation is more complicated for linearization inference. The Horvitz–Thompson estimator

$$\widehat{\text{var}}\left[\hat{T}_Y\right] = \sum_{i,j} \left(\frac{Y_i Y_j}{\pi_{ij}} - \frac{Y_i}{\pi_i} \frac{Y_j}{\pi_j} \right)$$

is still equivalent to an estimator that takes sums over the subpopulation

$$\widehat{\text{var}}\left[\hat{T}_Y\right] = \sum_{Z_i, Z_j > 0} \left(\frac{Y_i Y_j}{\pi_{ij}} - \frac{Y_i}{\pi_i} \frac{Y_j}{\pi_j} \right),$$

but the simplified computational formulas for special designs are not the same. For example, the formula for the variance of a total under simple random sampling (equation 2.2)

$$\text{var}\left[\hat{T}_X\right] = \frac{N-n}{N} \times N^2 \times \frac{\text{var}[X]}{n}$$

cannot be replaced by

$$\text{var}\left[\hat{T}_Y\right] \overset{?}{=} \frac{N-n}{N} \times N^2 \times \frac{\text{var}[X]}{n}$$

or even, defining n_D as the number sampled in the subpopulation, by

$$\text{var}\left[\hat{T}_Y\right] \overset{?}{=} \frac{N - n_D}{N} \times N^2 \times \frac{\text{var}[X]}{n_D}.$$

In order to use these simplified formulas it is necessary to work with the variable Y and use

$$\text{var}\left[\hat{T}_Y\right] = \frac{N-n}{N} \times N^2 \times \frac{\text{var}[Y]}{n}.$$

Operationally, this means that variance estimation in a subset of a survey design object in R needs to involve the $n - n_D$ zero contributions to an estimating equation. For two-phase designs, designs with calibration or post-stratification, or database-backed designs, this has done by setting the sampling weights to zero for observations outside the subpopulation. For simpler design objects, the observations outside the domain are discarded by subset(), and zero entries are added at the time of variance estimation.

A.3 TESTS IN CONTINGENCY TABLES

svychisq() computes the Wald statistic in a weighted least squares analysis, from Kochet al. [77], and the Pearson (score) statistic for a working loglinear model, from Rao and Scott [130]. For loglinear models, anova() computes the likelihood ratio (G^2) and score (X^2) statistics as described by Rao and Scott [130]. The procedure is to estimate the proportions in a multiway table of all variables and then treat these estimated proportions as if they came from a multinomial sample of the same size. Rao and Scott also give formulas for the standard errors of the coefficients in the loglinear model $(\hat{D}(\hat{\phi})$ in their section 2.3), and regTermTest() will compute Wald tests using these.

The Wald test statistics have an asymptotic χ^2_p distribution under the null hypothesis, but require relatively large sample sizes for this to be accurate. In regTermTest() the default is to use an F^p_d distribution, where

$$d = \text{nPSU} - \text{nStrata} + 1 - p.$$

The user can specify a different denominator degrees of freedom if desired.

The working score and working likelihood ratio tests do not have an asymptotic χ^2_p distribution. Their asymptotic null distribution is

$$G^2, X^2 \xrightarrow{d} \sum_{i=1}^{p} a_i \chi^2_1$$

where a_i are the limits of the eigenvalues of a "generalized design effect" matrix defined in Theorem 1 of [130]. For loglinear models, anova() returns the estimates of a_i in the $a component.

As the linear combination of χ^2_1 is not a standard distribution, approximations have usually been used to compute p-values. Two approximations were given by Rao and Scott [130]: a χ^2_p distribution scaled to have the right mean, and an F distribution with the right mean and variance. The F distribution, the "second-order Rao-Scott correction" is the default for both svychisq() and anova.svyloglin().

The asymptotic distribution can be computed by numerical integration. The characteristic function is

$$\phi(t) = \prod_{i=1}^{p} \phi_{\chi^2_1}(t)^a_i$$

where $\phi_{\chi^2_1}(t)$ is the characteristic function of a χ^2_1 variable. The distribution function can be recovered by integrating along the imaginary axis in the complex plane

$$F(x) = \frac{1}{2} - \int_{-\infty}^{\infty} \text{Im} \left(\frac{1}{2\pi t} \phi(t) e^{-ixt} \right) dt.$$

The integration is done with integrate(). In the right tail the integral converges slowly, so p-values below about 10^{-5} are not feasible in the current interpreted version of the code. Kuonen [86] gives a saddlepoint approximation for the asymptotic distribution. This is numerically unstable near $x = \sum a_i$, but is very accurate in the right tail. It is important to note that the numerical integration and saddlepoint approximation are effectively exact for the asymptotic distribution, not for the finite sample distribution. It is quite possible that approximations which are less accurate for the asymptotic distribution could be more accurate in finite samples. I am not aware of any simulation studies comparing the exact asymptotic distribution to other approximations in finite samples.

The function pchisqsum() produces the Satterthwaite approximation, the exact distribution by numerical integration, and the saddlepoint approximation. For method="integration" the code falls back to the Satterthwaite approximation for small p-values, and for method="saddlepoint" it falls back to the Satterthwaite approximation when $x < 1.05 \times \sum a_i$.

A.4 MULTIPLE IMPUTATION

Rubin's variance formula (equation 9.4) is strictly valid only for so-called *proper imputation*, where the imputations are created from the posterior distribution of a parametric Bayesian model. This involves fitting models

$$X_i \sim P(X_i | X_1, \ldots, X_{i-1}, X_{i+1}, \ldots, X_p; \theta)$$

to the distribution of each variable given all the others, then sampling $\tilde{\theta}$ from the posterior distribution of θ and then sampling X_i from the predictive distribution

$$P(X_i | X_1, \ldots, X_{i-1}, X_{i+1}, \ldots, X_p; \tilde{\theta})$$

with a different $\tilde{\theta}$ used for each imputed data set. An alternative way to create the imputed values is to fix θ at the maximum likelihood estimate $\hat{\theta}$ based on the observed data and then to sample X_i from

$$P(X_i | X_1, \ldots, X_{i-1}, X_{i+1}, \ldots, X_p; \hat{\theta})$$

for all the imputed data sets. This approach gives consistent estimates, and since the added uncertainty from sampling $\tilde{\theta}$ is not present these estimates are more precise than those from proper imputation. Unfortunately, equation 9.4 will not correctly estimate the variance (Wang and Robins [184], Robins and Wang [138]).

The correct variance for these "improper" imputations depends on the model used for imputation and so is not available when public-use data sets are imputed using more detailed microdata. When the proportion of missing data is small the bias in equation 9.4 is likely to be small.

Kim et al. [74] describe a bias in the multiple-imputation standard errors for subpopulation estimation. If a proper multiple imputation were carried out separately in the subpopulation, the standard error estimates would be correct, although the inference would be inefficient. When the multiple imputation is carried out using the entire data set (and under the assumption that the model is correct) the model parameters will be estimated more accurately and the uncertainty will be smaller than if only the subpopulation were used. In the extreme case of a very small domain and very accurately imputed variable, the standard error of a simple subpopulation mean could be smaller with the data missing than with the data observed. Rubin's variance formula does not take this extra information into account, and so overestimates the standard errors. This bias is of a different kind from that described by Wang and Robins, and can perhaps be defended as a protection against model misspecification: the extra precision comes from assuming that the model fitted to the entire population still works for small subpopulations, and this precision gain is not reliable in practice. It would be an interesting research project to see how the calculations of Kim et al. change if the imputation model is very slightly misspecified.

Shao and Sitter [162] describe a combination bootstrap and imputation technique that gives correct standard errors from any form of imputation, and so gives correct inference from both proper and improper multiple imputation. The technique involves performing imputation separately on each bootstrap sample, and so would generate quite large data sets and require considerable computation, but not enough to make it infeasible with modern computers. I am not aware of any surveys that currently use this technique, but it is certainly a promising future approach.

A.5 CALIBRATION AND INFLUENCE FUNCTIONS

Särndal [149], following Estevao and Särndal [46], gives the example of estimating the mean of a variable Y in a subpopulation, where membership in the subpopulation of interest is not known for the population, but membership in an overlapping subpopulation is known.

Suppose we are interested in estimating the total of Y over a subpopulation $\mathcal{D}$. Without auxiliary information the estimator is

$$\hat{T}_{\mathcal{D}} = \sum_{i \in \mathcal{D}, R_i = 1} \frac{1}{\pi_i} Y_i = \sum_{R_i = 1} \frac{1}{\pi_i} D_i Y_i$$

where D is the indicator variable for membership in $\mathcal{D}$. If auxiliary variables X were available and the population total for XD were known, an improved regression estimator would be

$$\hat{T}_{\mathcal{D}}^{(\text{reg})} = \sum_{R_i = 1} \frac{1}{\pi_i} D_i (Y_i - X_i \hat{\beta}) + \sum_{i=1}^{N} D_i X_i \hat{\beta}$$

where $\hat{\beta}$ could be estimated by a regression of Y on X over the sampled members of the subpopulation or (trading bias and variance) by a regression over the whole sample. The regression estimator using just the subpopulation is exactly the same as a calibration estimator using XD as auxiliary variables.

Suppose, however, that population data is not available on membership in $\mathcal{D}$ but is available for a closely related subpopulation $\mathcal{D}^*$ (perhaps overlapping, perhaps a subset). Let D_i^* be the indicator variable for membership in $\mathcal{D}^*$ and suppose that the population total is known for XD_i^*.

Estevao and Särndal argued that generalizing the regression estimator to this problem would give an estimator

$$
\begin{aligned}
\hat{T}_{\mathcal{D}}^{*(\text{reg})} &= \sum_{R_i=1} \frac{1}{\pi_i} D_i Y_i + \left(\sum_{i=1}^{N} D_i^* X_i - \sum_{R_i=1} \frac{1}{\pi_i} D_i^* X_i \right) \hat{\beta} \\
&= \sum_{R_i=1} \frac{1}{\pi_i} \left(D_i Y_i - D^* X_i \hat{\beta} \right) + \sum_{i=1}^{N} D_i^* X_i \hat{\beta}
\end{aligned}
$$

where $\hat{\beta}$ might be estimated on $\mathcal{D}^* \cap \mathcal{D}$ or, attempting to borrow strength, on all of $\mathcal{D}^*$.

Their calibration approach uses XD_i^* as auxiliary variables to give an estimator

$$
\hat{T}_{\mathcal{D}}^{*(\text{cal})} = \sum_{R_i=1} \frac{g_i}{\pi_i} D_i Y_i .
$$

Estevao and Särndal show that $\hat{T}_{\mathcal{D}}^{*(\text{cal})}$ and $\hat{T}_{\mathcal{D}}^{*(\text{reg})}$ are not the same and that $\hat{T}_{\mathcal{D}}^{*(\text{reg})}$ is more efficient.

In fact, the more efficient estimator $\hat{T}_{\mathcal{D}}^{*(\text{reg})}$ can be constructed as a regression estimator by working with influence functions rather than with Y. The population estimating equation for $T_{\mathcal{D}}$ is

$$
T_{\mathcal{D}} \sum_{i=1}^{N} \frac{1}{\pi_i} D_i Y_i
$$

so the influence functions are $Y_i D_i - T_{\mathcal{D}}/N$. These influence functions suggest a regression with $Y_i D_i$ as the outcome, rather than Y_i itself. The regression estimator is then

$$
\hat{T}_{\mathcal{D}}^{*(\text{infl})} = \sum_{R_i=1} \frac{1}{\pi_i} (D_i Y_i - D_i^* X_i \hat{\gamma}) + \sum_{i=1}^{N} D_i^* X_i \hat{\gamma}
$$

which is very similar in form to $\hat{T}_{\mathcal{D}}^{*(\text{reg})}$, the difference being that that the coefficients $\hat{\gamma}$ are from a regression of DY on D^*X rather than the regression of Y on X used for $\hat{\beta}$.

Since $\hat{T}_{\mathcal{D}}^{*(\text{infl})}$ is just a regression estimator of a population total, it is equivalent to the calibration estimator of the population total with the same auxiliary variables, that is, to $\hat{T}_{\mathcal{D}}^{*(\text{cal})}$. "Regression thinking" combined with "influence function thinking" gives the same estimators as "calibration thinking."

A.6 CALIBRATION IN RANDOMIZED TRIALS AND ANCOVA

Consider a two-group randomized trial, in which a baseline variable X is measured, $n/2$ participants are randomized to each of treatments A or B, and then an outcome

Y is measured. The summary of interest is the average causal effect of treatment on Y. This can be estimated either as the difference in mean of Y between the treatment groups or as the coefficient of a treatment term in a regression model for Y: if treatment is coded $Z = +1$ for A and $Z = -1$ for B we have

$$E[Y|Z] = \mu + Z\delta$$

where $\delta/2$ is the average causal effect of randomization to treatment. The obvious and standard estimator of $\delta/2$ is the difference in means between treatment A and treatment B

$$\hat{\delta}/2 = \frac{1}{N/2} \sum_{Z_i=1} Y_i - \frac{1}{N/2} \sum_{Z_i=-1} Y_i.$$

Using the potential-outcomes formulation of causation we consider the randomized trial as a sample from a finite population. In the finite population, each participant i has two potential outcomes: $Y_{(A)i}$ if assigned treatment A and $Y_{(B)i}$ if assigned treatment B. The randomization process samples one potential outcome for each participant. The use of randomization to assign treatments guarantees that the sampling probabilities are independent of the potential outcomes and of X. Under 1:1 randomization these sampling probabilities at the second phase are $1/2$ for each potential outcome. The observed value of Y is the one for the assigned treatment Z_i, namely $Y_i = Y_{(z_i)i}$.

The treatment effect for an individual is $Y_{(A)i} - Y_{(B)i} = \sum_z Y_{(z)i} Z_i$, so the average treatment effect is the population mean of YZ, where the expectation is taken over the two treatments and potential outcomes for each individual. The IPTW estimator $\hat{\delta}_{IPTW}$ of δ is a probability-weighted sum of YZ over the observed outcomes and treatments. This reduces to the group difference in means

$$
\begin{aligned}
\hat{\delta}_{IPTW}/2 &= \frac{1}{n} \sum_{i=1}^{n} \frac{1}{\pi_i} Y_i Z_i \\
&= \frac{1}{n} \sum_{i=1}^{n} \frac{1}{1/2} Y_i Z_i \\
&= \frac{1}{n/2} \left(\sum_{Z_i=1} Y_i - \sum_{Z_i=-1} Y_i \right) \\
&= \hat{\delta}/2.
\end{aligned}
$$

When additional baseline variables X are available the treatment effect can be estimated by a regression of Y on X and Z, fitting the model

$$E[Y|Z, X] = \mu + Z\delta + X\beta. \tag{A.4}$$

For notational simplicity we consider only univariate X, but exactly the same arguments apply for multivariate X. The baseline-adjusted estimator $\hat{\delta}_{reg}$ satisfies the

Normal equations:

$$\sum_{i=1}^{n}(Y_i - \mu - Z_i\delta - X_i\beta) = 0$$

$$\sum_{i=1}^{n}Z_i(Y_i - \mu - Z_i\delta - X_i\beta) = 0$$

$$\sum_{i=1}^{n}X_i(Y_i - \mu - Z_i\delta - X_i\beta) = 0. \qquad (A.5)$$

This estimator $\hat{\delta}_{reg}$ is more efficient than the difference in means between treatment groups. If X and Y are highly correlated the efficiency gain can be large. Because Z is randomly assigned and independent of X the regression estimator is unbiased for δ regardless of whether the regression is correctly specified; a misspecified model just leads to a smaller gain in efficiency.

An alternative way to use baseline variables X is by calibrating the sampling weights. In the first-phase sample each individual appears once in each treatment group, so the sample mean of XZ is identically zero. In the observed sample there will be small imbalances in X between treatment groups, so that the IPTW estimator of the mean of XZ is not exactly zero.

When we calibrate on XZ to the first-phase sample the calibration constraints (7.3) are

$$\sum_{i=1}^{N}\frac{g_i}{1/2}X_i Z_i = \sum_{i:Z_i=1} 2g_i X_i - \sum_{i:Z_i=-1} 2g_i X_i = 0,$$

i.e., perfect balance in the mean of X across treatment groups. In fact, we will calibrate to $S = (XZ, Z, 1)$, where the calibration on $(1, Z)$ ensures that the sum of the weights stays equal to $2N$ and the mean of Z stays equal to zero. These additional conditions result in the calibrated estimator being algebraically equal to $\hat{\delta}_{reg}$, calibrating only on XZ gives an asymptotically equivalent estimator. We will use the equivalence between the calibration estimator and the survey regression estimator in equation 7.2. We write $(\alpha_0, \alpha_1, \alpha_2)$ for the regression coefficients in equation 7.2. These satisfy the weighted least-squares equations

$$\frac{1}{1/2}\sum_{i}(Y_i Z_i - \alpha_0 - \alpha_1 Z_i - \alpha_2 Z_i X_i)1 = 0$$

$$\frac{1}{1/2}\sum_{i}(Y_i Z_i - \alpha_0 - \alpha_1 Z_i - \alpha_2 Z_i X_i)Z_i = 0$$

$$\frac{1}{1/2}\sum_{i}(Y_i Z_i - \alpha_0 - \alpha_1 Z_i - \alpha_2 Z_i X_i)X_i Z_i = 0.$$

Using the fact that $Z_i = 1/Z_i$ and $Z_i^2 = 1$, we can rewrite this as

$$\sum_i (Y_i - \alpha_0 Z_i - \alpha_1 - \alpha_2 X_i) Z_i = 0$$

$$\sum_i (Y_i - \alpha_0 Z_i - \alpha_1 - \alpha_2 X_i) 1 = 0$$

$$\sum_i (Y_i - \alpha_0 Z_i - \alpha_1 - \alpha_2 X_i) X_i = 0$$

which are the least-squares equations for the model in equation A.5, with $\alpha_0 = \delta$, $\alpha_1 = \mu$, and $\alpha_2 = \beta$, so $\hat{\alpha}_0 = \hat{\delta}_{reg}$.

According to equation 7.2, the calibration estimator for the total of YZ is

$$N\hat{\delta}_{cal} = \frac{1}{1/2} \sum_i (Y_i Z_i - \hat{\alpha}_0 - \hat{\alpha}_1 Z_i - \hat{\alpha}_2 Z_i X_i) + T$$

where T is the population total of the predicted values. The first term is zero, from the definition of α, an unusual special case that occurs because the sampling weights are constant. The second term T expands to

$$T = \frac{1}{1/2} \sum_i (\hat{\alpha}_0 + \hat{\alpha}_1 Z_i + \hat{\alpha}_2 Z_i X_i).$$

Since the sums of $Z_i X_i$ and Z_i are identically zero over the potential-outcome population, T simplifies to

$$T = N\hat{\alpha}_0 = N\hat{\delta}_{reg}$$

and so $\hat{\delta}_{cal} = \hat{\delta}_{reg}$.

Appendix B

BASIC R

B.1 READING DATA

B.1.1 Plain text data

Section 1.4 described how to read comma-separated text (`csv`) files. Other plain text formats can be read with the function `read.table()`. This function is more flexible, so more options need to be specified. The PBC data set is stored with spaces between the columns, and with variable names in the first line of the file. It can be read with the code

```
pbc <- read.table("pbc.txt", header=TRUE)
```

where the option `header=TRUE` specifies that the first line of the file contains variable names.

If a file appears to have columns separated by spaces but does not read correctly it is likely that it actually has columns separated by tab characters. These can be read using

*

Complex Surveys: A Guide to Analysis Using R. By Thomas Lumley
Copyright © 2010 John Wiley & Sons, Inc.

231

```
dataset <- read.table("tabseparated.txt", header=TRUE, sep="\t")
```

where \t specifies a tab character. In the same way, comma-separated data could be read with read.table() using sep="," to specify a comma as the separator. Another cause of difficulties is the varying ways that missing data are represented in the file. By default, read.table() assumes that missing data will be coded NA. The coding for missing data can be specified with the na.strings option. For example, if missing observations are indicated by an asterisk

```
dataset <- read.table("tabseparated.txt", header=TRUE, sep="\t",
    na.strings="*")
```

There are many more options to read.table, these are described in the help page, which can be obtained with help(read.table) or simply ?read.table.. The help pages can be searched with help.search(), so that help.search("read") lists 37 functions with "read" in their name or help page title. For more general searches including help pages for all packages on CRAN and messages on the mailing list, RSiteSearch() sends a search query to http://search.r-project.org (if you are connected to the internet).

Other types of plain text data include fixed-width format text data without any separation between the variables, and XML. Fixed format data can be read in using read.fwf(), specifying the starting and ending positions for each variable, or read.fortran(), specifying variable types and widths. XML can be read with the XML package.

B.2 DATA MANIPULATION

B.2.1 Merging

Data from a survey often come in multiple files that need to be merged for analysis. If the files are very large it may be better to use a database program to do the merging, as described in section D, but for data sets of moderate size the R function merge() is convenient. Figure B.1 shows how to read and merge the NHANES 2003–2004 demographic and blood pressure data sets. The two data sets are given as the first two arguments to merge(), the by="SEQN" argument says that SEQN is the name of a unique identifier variable in both data sets, and all=FALSE says that individuals who have a record in only one of the data sets should not be included.

```
demo <-read.xport("~/nhanes/demo_c.xpt")
bp <-read.xport("~/nhanes/bpx_c.xpt")
both <- merge(demo, bp, by="SEQN", all=FALSE)
```

Figure B.1 Merging the demographic and blood pressure data sets

The use of all=FALSE depends on whether the missing records reflect non-response or a different subsample for some variables. The blood pressure data are available only for participants in the clinical examination sample, and will be analyzed using the sampling weights WTMEC2YR for the clinical examination, so only those individuals should be in the merged data set.

B.2.2 Factors

Categorical variables are represented in R as *factors*. This is superficially similar to the use of value labels in many other packages, but in a factor variable the labels, rather than the underlying numeric values, are used for further computations. A factor can be defined with the factor() function, e.g.,

```
demo$sex <- factor(demo$RIAGENDR, levels=1:2,
            labels=c("Male","Female"))
```

Factors can also be constructed from continuous variables with cut(), eg

```
demo$agegrp <- cut(demo$RIDAGEYR, c(0,21,65,Inf))
```

creates a factor with levels (0,21], (21,65] and (65,Inf], where Inf, meaning "infinity" is a convenient way to specify no upper limit. By default the right endpoint is included in the category, there is an option right=FALSE to change this.

A factor variable cannot be used in arithmetic expressions, except for tests of equality. Equality is based on the labels, not on the numeric codes, so the male participants could be selected by

```
men <- subset(demo, sex=="Male")
```

When a factor is used in a regression model it is automatically coded as a set of indicator variables, with the first level omitted. The function relevel() can be used to change which level is first, to produce a different reference level. Note that relevel changes the numeric codes, but keeps the same label for each observation; this highlights the difference between factors and variable labels in many other packages.

B.3 RANDOMNESS

The sample() function takes samples with and without replacement from its input, so sample(x, 10) takes a sample of 10 elements without replacement from x. There is a prob argument to supply sampling probabilities, but note that when sampling without replacement this does sequential sampling, not true PPS sampling. True PPS sampling requires an additional package such as sampling [173].

There are functions to sample from a wide range of probability distributions. Some of these are

rnorm() Normal distribution

rbinom() Binomial distribution

`rt()` *t* distribution

`rchisq()` χ^2 distribution

`rf()` *F* distribution

`runif()` uniform distribution

The first argument for each of these functions is the number of random numbers to generate, subsequent arguments give any necessary parameters, such as the mean and standard deviation for a Normal distribution.

Supporting `rnorm()` for sampling from a Normal distribution are `pnorm()` for the Normal cumulative distribution function, `qnorm()` for its inverse, the Normal quantile or tail function, and `dnorm()`, the Normal probability density function. All the probability distributions supported by R have these four functions available, and their names are obtained by changing the first letter of the random sampling function, just as for the Normal.

To get reproducible results from a random process such as a simulation it is necessary to be able to reproduce the stream of random numbers. The function `set.seed()` takes a single integer argument and uses it to restart the random number stream. In the future, calling `set.seed()` with the same argument will restart the random number stream in the same place. Examples of using `set.seed()` can be seen in Chapter 8.

B.4 METHODS AND OBJECTS

Functions such as `subset()` and `summary()` are *generic functions*, with different behavior depending on the type of argument they are given. The `subset()` function is a placeholder

```
> subset
function (x, ...)
UseMethod("subset")
```

that tells R to find the appropriate specialized function, or *method* to match the arguments it has been given. The available methods can be displayed with the `methods()` function

```
> methods("subset")
 [1] subset.DBIsvydesign*         subset.ODBCsvydesign*
 [3] subset.data.frame            subset.default
 [5] subset.matrix                subset.survey.design*
 [7] subset.svyDBimputationList*  subset.svyimputationList*
 [9] subset.svyrep.design*        subset.twophase*
```

The part of the name after `subset` is the *class* of argument that the method can handle, with the `subset.default` method being called if there is no more appropriate

```
a.median <- function(x, na.rm = TRUE){
  if (na.rm){
    x <- x[!is.na(x)]
  } else {
    if(any(is.na(x))) return(NA)
  }
  x <- sort(x)
  n <- length(x)
  if (n %% 2 == 1)
      x[(n+1)/2]
  else
      (x[n/2]+x[n/2+1])/2
}
```

Figure B.2   A simple function: computing a median

specialized method. The class of an argument is reported with the `class()` function, for example, `class(demo)` reports that the NHANES 2003–2004 demographics data set has class `"data.frame"`.

The help page for a generic function may include help for some methods, but it is often necessary to request help specifically for the method, eg `?subset.twophase` for subsets of two-phase survey design objects. Most of the functions in the survey package are generic, with methods that depend on how the data are stored and on what sort of design information is included with the survey.

B.5 ★ WRITING FUNCTIONS

Both data analysis and programming in R involve writing your own functions to package up pieces of code so you can reuse them later. Figure B.2 shows a simple function that computes a median. The first line specifies that we are defining a function, which will be called `a.median` and which will have two arguments. The second argument, `na.rm`, has a default value of `TRUE`. If the user does not specify the `na.rm` argument the default value will be used. The function begins by handling missing data. If `na.rm` is `TRUE`, the missing data are removed: `is.na()` gives `TRUE` for missing and `FALSE` for non-missing, and the exclamation point means "not", so the vector x is replaced by just the non-missing elements. If `na.rm` is `FALSE` we test whether there are any NA values. If so, the median is unknown and the function simply gives up and returns `NA`.

The remaining code computes the median. It can ignore the possibility of missing data, since at this point there cannot be any. The median will be the middle element, if the sample size is odd, or the average of the middle two, if the sample size is even. The vector x is sorted into increasing order, so that the middle element will be in the middle position. The `%%` operator gives the remainder when n is divided by 2:

1 if n is odd, 0 if n is even. If n is odd the middle element is at position (n+1)/2. If n is even the middle two elements are at positions n/2 and n/2+1. A return() statement is not needed (and is typically not used) because the last value computed in the function will be returned as the value. Variables created (such as n) or modified (such as x) inside the function are local to the function; modifications that happen inside a function cannot affect variables outside.

The built-in R function to compute the median is median.default. It is very similar to Figure B.2, but has a few extra features. It will correctly handle the situation where x has length zero (after removing missing values) and it uses a slightly faster "partial sort" that does not sort the vector x completely into order, but stops when the middle elements are in the right places.

B.5.1   Repetition

R has facilities for loops that are similar to those in other programming languages, but also has a collection of tools for calling a function repeatedly.

replicate() executes the same piece of code many times and collects the results. This is obviously only useful when the code does something different each time it is run, in particular, for simulations. For example, Figure 2.1 was created using replicate() together with sample() to take simple random samples. Use set.seed() to start the sequence of random numbers in a reproducible way.

lapply(), sapply() call a function using each element of a list or vector in turn as the argument. Many loops can be rewritten using these functions.

apply() calls a function using each row or column of a data set in turn as the argument. Figure 8.2 uses apply() to compute means and standard errors from simulation results.

by(), tapply() call a function on subsets of a given vector or data set.

Example: Consider the problem of computing a table of means and standard deviations for several variables, separately for groups of people, e.g., multiple blood pressure measurements broken down by age and sex. Ignoring the sampling design for the sake of a simpler example, we can use the data set both that is created in Figure B.1. The variables BPXSY1, BPXDI1, to BPXSY3, BPXDI3 are three sets of blood pressure measurements. Figure B.3 shows how to compute means and standard deviations by age and sex, building up the analysis from smaller pieces.

Computing the mean of each variable can be done with apply(). The first argument is the data set to analyze, the second is 1 for rows or 2 for columns, and the third is the analysis function, in this case mean(). Any remaining arguments (in this case na.rm) are passed to the analysis function. The second call to apply() is the same except that the analysis function is sd(), to compute the standard deviation of each column of the data set.

```
> bpmeas <- both[, c(47,48,50,51,53,54)]
> apply(bpmeas, 2, mean, na.rm=TRUE)
   BPXSY1    BPXDI1    BPXSY2    BPXDI2    BPXSY3    BPXDI3
119.87313  65.06886 119.11801  65.85144 118.39943  65.76493

> apply(bpmeas, 2, sd, na.rm=TRUE)
  BPXSY1   BPXDI1   BPXSY2   BPXDI2   BPXSY3   BPXDI3
20.64478 14.87837 19.82983 13.91242 19.33024 13.98752

> apply(bpmeas, 2,
      function(x) c(m=mean(x, na.rm=TRUE), s=sd(x, na.rm=TRUE)))
      BPXSY1    BPXDI1    BPXSY2    BPXDI2    BPXSY3    BPXDI3
m 119.87313  65.06886 119.11801  65.85144 118.39943 65.76493
s   20.64478 14.87837  19.82983 13.91242  19.33024 13.98752

> by(bpmeas$BPXSY1, list(sex=both$RIAGENDR), mean, na.rm=TRUE)
sex: 1
[1] 121.2939
------------------------------------------------------------
sex: 2
[1] 118.4943

> both$agegp<-cut(both$RIDAGEYR,c(0,30,50,70,85))
> by(bpmeas, list(sex=both$RIAGENDR,agepg=both$agegp),
    function(subset) apply(subset, 2,
        function(x) c(m=mean(x, na.rm=TRUE), s=sd(x, na.rm=TRUE))
        )
     )
sex: 1
agepg: (0,30]
      BPXSY1    BPXDI1    BPXSY2    BPXDI2    BPXSY3    BPXDI3
m 112.71383  59.44296 112.63345  60.76584 112.13275 60.85631
s   11.46825 14.22283  11.20644 13.09770  11.28567 13.29416
------------------------------------------------------------
sex: 2
agepg: (0,30]
      BPXSY1    BPXDI1    BPXSY2    BPXDI2    BPXSY3    BPXDI3
m 106.66012  59.42331 106.410397 60.26813 105.942416 60.42275
s   10.21637 11.71615   9.979094 11.89340   9.984362 11.95204
... output truncated ...]
```

Figure B.3   Using by() and apply()

The third call to apply() uses a new function that is created inside the call. The code

```
function(x) c(m=mean(x, na.rm=TRUE), s=sd(x, na.rm=TRUE))
```

specifies a function that takes one argument (which it calls x). The function computes the mean and standard deviation of this x and composes them into a single vector with names m and s. This vector is then returned as the function value. The call to apply() applies this function to each variable in the data set, so the final result is the mean and standard deviation of each variable, in a 2×6 table.

Now we know how to compute statistics for each of a set of variables, the next step is to compute separately for subsets of the data. This is done with by(). As with apply(), the first argument is a vector or data set to analyze and the third argument is an analysis function. The second argument is different. It is a list of grouping variables that specify the subsets to analyze. The first call to by() in Figure B.3 uses mean() as the analysis function, and the variable RIAGENDR as the grouping variable to compute the mean of BPXSY1 for men and women.

The second call to by() is more complicated. The first argument, giving the data to be analyzed, is the entire set of blood pressure measurements. The second argument gives the grouping variables. There are now two of these, sex and an age group created from RIDAGEYR using cut().. The third argument, the analysis function, is the entire call to apply() that was used two lines earlier, wrapped in a new function. The analysis function takes one argument, which it then calls subset. It then computes the mean and standard deviation of each column of subset and returns the result in a 2×6 table. As this analysis function is called from by(), the data passed to it will be the subsets of the full data set corresponding to the grouping variable: the analysis function will first be called with all the data for men under age 30, then with the data for women under age 30, then with the data for men aged 30–50, and so on. The output now gives the 2×6 table of means and standard deviations for each sex/age combination. Only the first two combinations are shown, for reasons of space.

B.5.2 Strings

Character strings can be concatenated with paste(), so that

```
paste("imputation",1:5,".dat", sep="")
```

creates a vector of five file names, imputation1.dat to imputation5.txt. The function can also be used to collapse a vector of strings, so that paste(vars, collapse=", ") turns a vector of variable names into a single comma-separated string. Substitution inside a string is done with sub(), so

```
sub("@", i, "C:/boring/data/file@.txt")
```

will substitute the value of the variable i into the file name. Vectors of strings can be searched with grep(), eg, grep("wt",names(dataset)) to find out which variable names have the substring wt.

Appendix C

COMPUTATIONAL DETAILS

C.1 LINEARIZATION

All standard error computations for linearization go through svyrecvar(), which is passed either the estimating functions or the influence functions for the estimator. The first step is to project the influence functions orthogonal to any post-stratifying or raking or calibration variables. For post-stratification, post-stratum means are computed using the original sampling weights and subtracted off. For raking, the post-stratification procedure is done iteratively. For calibration, a weighted QR decomposition of the matrix of auxiliary variables weighted by the original sampling weights is used to compute residuals. The resulting residuals are passed to multistage(), which implements the recursive algorithm for multistage variances.

The variance contribution for the current stage is computed by onestage(), which calls onestrat() to do the actual calculations for one stratum, and then multistage() is recursively called for the next stage of sampling. In onestrat() the observations for each cluster are added, then centered within each stratum and

*

Complex Surveys: A Guide to Analysis Using R. By Thomas Lumley
Copyright © 2010 John Wiley & Sons, Inc.

the outer product is taken of the row vector resulting for each cluster. This is added within strata, multiplied by a degrees-of-freedom correction and by a finite population correction (if supplied), and added across strata in `onestage()`.

If there are fewer clusters (PSUs) in a stratum than in the original design, extra rows of zeroes are added to allow the correct subpopulation variance to be computed.

C.1.1   Generalized linear models and expected information

The derivative of the estimating functions used in linearization variance calculations for `svyglm()` comes from the information matrix computed by `glm()`. This is the *expected* value of the second derivative matrix under the working model, not the observed value. For the canonical link functions, which includes logistic and linear regression and Poisson regression, the observed and expected derivatives are identical. For other link functions, including relative risk regression, the observed and expected derivatives are not identical, but they are typically very close unless there are very strong associations present and the model is so egregiously misspecified as to be useless as a population summary.

C.2   REPLICATE WEIGHTS

C.2.1   Choice of estimators

The computations to estimate the variance from the replicates are done by `svrVar()`. The variance is computed from squared deviations between the replicates and their mean, not deviations between the replicates and the point estimate. That is, if the m replicate estimates are T_i^* and their mean is $\bar{T}^*$, the variance estimate is

$$\widehat{\text{var}}[\hat{T}] = \texttt{scale} \times \sum_{i=1}^{m} (T_i^* - \bar{T}^*)^2 \times \texttt{rscales}_i$$

where `scale` and `rscales` are the arguments to `svrepdesign()`. The replicate mean $\bar{\theta}^*$ is returned as an attribute of the result so that the alternative estimator

$$\widehat{\text{var}}[\hat{T}] = \texttt{scale} \times \sum_{i=1}^{m} (T_i^* - \hat{T})^2 \times \texttt{rscales}_i$$

can be computed by adding

$$\texttt{scale} \times (\bar{T}^* - \hat{T})^2 \times \sum_{i=1}^{m} \texttt{rscales}_i.$$

Any replicates that are `NA` or `NaN` are dropped, so variances will then tend to be biased downwards. A warning is given and the number of missing values is also returned as an attribute of the result.

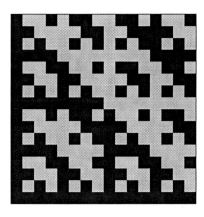

Figure C.1 Hadamard matrix of dimension 16

C.2.2 Hadamard matrices

The problem of finding sets of weights with full orthogonal balance is not trivial, but fortunately is of interest in other areas of mathematics. A Hadamard matrix is a matrix with entries ± 1 and columns orthogonal to each other, these matrices have largest possible determinant given that their elements are bounded by 1. Adding 1 to each element of a Hadamard matrix gives a set of BRR weight multipliers.

The number of columns, n, of a Hadamard matrix must be a multiple of four; it is conjectured that they exist for all multiples of four, and the first n for which no matrix is currently known is 668. More usefully for the survey package, there are two simple constructions that give Hadamard matrices for most sizes [26]. Given any Hadamard matrix H, a matrix with twice the dimension can be constructed by pasting together three copies of H and one of $-H$

$$H_{2n} = \begin{pmatrix} H_n & H_n \\ H_n & -H_n \end{pmatrix}. \qquad (C.1)$$

When $n = p + 1$ for a prime p a Hadamard matrix can easily be computed by the Paley construction. Combining these gives matrices of dimension $2^k(p + 1)$. In addition, the package stores a few matrices that do not come from these constructions, notably $n = 28$ and $n = 36$. Figure C.1 shows a Hadamard matrix of dimension $16 = 2^1 \times (7 + 1)$. The diagonal stripes are characteristic of the Paley construction and the repeated blocks show how the size has been doubled. The Paley construction actually allows for matrices of dimension $p^r + 1$ for any integer power r, but implementing this more general construction requires a convenient way of finding a representation for the field $\mathbb{Z}_{p^r}$, e.g., as polynomials over $\mathbb{Z}_p$, which doesn't seem as easy to automate.

C.3 SCATTERPLOT SMOOTHERS

The default in svysmooth(), method = "locpoly", uses a local linear regression smoother with Gaussian kernel weights, as implemented in the KernSmooth package [182]. The data are binned to 401 cells before smoothing, for speed. When used for density estimation the data are binned to 401 cells and the local linear regression smoother is applied to the resulting histogram. The bandwidth argument specifies the standard deviation of the Gaussian kernel.

The regression quantile estimators fit quantile linear regression models using the rq() function in the quantreg package [80]. The predictors are natural cubic regression splines with knots at evenly spaced quantiles of the data, with the number of internal knots being one less than the number of degrees of freedom specified by the df argument. The quantreg package has a function rqss() that fits smoothing splines by quantile regression, but at the time of writing this does not accept weights. It is likely that this will replace the regression splines in some future version of the survey package.

C.4 QUANTILES

Estimation of quantiles and their standard errors is relatively difficult and is avoided by many packages. The quantiles computed by svyquantile() differ from those computed by SUDAAN for at least two reasons.

The first issue is the definition of the quantile. Consider a data set with observations 1, 1, 1, 2, 2, 2, 2, 2, 3, with equal sampling weights. It is possible either to treat the variable as discrete or as a rounded version of a continuous variable. If it is treated as discrete the sample median is exactly 2. If it is treated as a rounded continuous variable the median should be between 1 and 2. For example, this data set would be a plausible realization from a $N(1.5, 1)$ distribution, which has median 1.5. R treats the variable as discrete and gives 2 as the estimated median. According to the online FAQ, SUDAAN will interpolate, and if I understand the FAQ correctly it will give 1.29 as the estimated median. Either of these approaches is reasonable and it is easy to find scenarios where either one would be preferable. The true standard error of the "continuous" estimator will be larger than the true standard error of the "discrete" estimator.

Estimating standard errors is also challenging. Both R and SUDAAN (Shah and Vaish [161]) use Woodruff's method as the default [189]. This involves treating the estimated median as a fixed constant, computing a confidence interval for the proportion exceeding this constant, and finding the quantiles corresponding to the endpoints of this confidence interval. Figure C.2 shows this procedure for the total monthly household income variable from wave 1 of the 1996 SIPP panel. Drawing a horizontal line at a quantile of 0.5 and then following down to the x-axis when the line meets the cumulative distribution gives the estimated median income as \$2648/month. Using svymean() we can then estimate a confidence interval for the proportion of incomes below \$2648/month. The standard error is 0.00305, so the

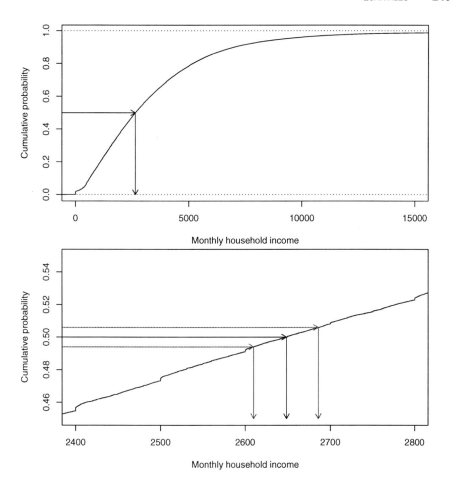

Figure C.2   Estimating the median and a confidence interval from the cumulative distribution function for monthly household income

confidence interval is $(0.5 - 1.96 \times 0.00305, 0.5 + 1.96 \times 0.00305)$, indicated by the gray horizontal arrows on the lower panel of Figure C.2. From where these arrows meet the cumulative distribution curve, we follow the vertical arrows to the x-axis to obtain the confidence interval for median monthly income, ($2610, $2686). The curves in Figure C.2 are fairly smooth, apart from small jumps at multiples of $100, because income is recorded to the nearest dollar. The uncertainty in the median is much larger than the size of the jumps, so they do not affect inference. If the data were recorded to the nearest $100 the cumulative distribution curves would be much more jagged and it might be unclear where the arrows should meet the curve. In this case, however, the ambiguity would be of about the same size as the rounding errors in the data and so should typically not be very important in practical data analysis.

To obtain an approximate standard error from the confidence interval, the length of the interval is divided by 2×1.96. In this example the standard error estimate is

$$(\$2686 - \$2610)/(2 \times 1.96) = \$19.4.$$

Although the same basic algorithm is used, the confidence interval for the proportion is computed differently by R and SUDAAN. R uses a Normal distribution and SUDAAN uses the exact binomial distribution for an estimated "effective sample size" (Korn and Graubard [81]). The end result is that the quantile estimates are similar but not identical, with the R estimate more likely to fall exactly on an observed value. The true standard error will be lower for the R estimate, and in addition there are differences in estimating the standard errors. The agreement will be closer if the data are truly continuous, without ties. SUDAAN will report wider confidence intervals because it is attempting to cover the population median of the underlying continuous variable and R is only attempting to cover the population median of the recorded data variable. The confidence intervals in R are probably too asymmetric for extreme quantiles because of the Normal approximation used in Woodruff's method. Future versions of the **survey** package will provide options for tie-handling and the exact binomial approach of Korn and Graubard [81].

C.5   BUG REPORTS AND FEATURE REQUESTS

The **survey** package, like all software except possibly TEX, contains bugs. It is probably also missing features that some users need, and is inefficient in some calculations. Reports of problems and requests for new features are always welcome. Please remember that if you have some serious problem with the **survey** package it is because I don't know that the problem exists, so it may be difficult to track it down based just on a complaint. It is helpful to give as much detail as you can about what you did, what happened, and how this differed from what you expected. It is possible that the package is doing exactly what I wanted it to do, in which case it will be necessary for us to discuss whether I am wrong.

The most useful requests are ones that allow me to have the same problem you have. If there is a bug or a usability problem the best possible solution is to send me the data set and script that has the problem. Since survey data are often large and often confidential this may be impossible. You may be able to produce a simpler example that has the same problem: this is just as good. With a real example I can use the full debugging and tracing facilities of R. If you can't send me a data set, please at least try to see if the problem occurs with the most recent version of the **survey** package — trying to track down a bug when there is no reproducible example and it doesn't even exist anymore will be needlessly frustrating for both of us.

For *feature requests* I need a reference (that I have access to) giving sufficient detail about the method to allow implementation, and, ideally a data set where the correct result is known, for testing.

Appendix D

Database-backed design objects

D.1 LARGE DATA

R has well-known difficulties in handling large data sets on small computers: it is recommended that a data set be no larger than 1/3 of memory, and that on a 32-bit system a data set should be no larger than 10% of the address space or about 300Mb regardless of the size of memory. Data files from national surveys can easily exceed these limits. One solution is to use a more powerful computer: 64-bit Linux servers with 16 Gb of memory are available in the US for about $2500 at the time of writing and can handle almost any complex survey design. Making major changes in one's computing environment may difficult, however, and it is useful to have a solution that applies to any hardware and operating system.

Rather than loading the entire data set into memory, it is possible to store the data in a relational database and give the information about where the data is located in the svydesign() call. Only the variables used in a particular computation will be loaded into memory, and only for the duration of that computation. Because less

*

Complex Surveys: A Guide to Analysis Using R. By Thomas Lumley
Copyright © 2010 John Wiley & Sons, Inc.

memory is used, computations are likely to be faster for large data sets, although they will be slower for small data sets because of the need to repeatedly load data from the database files. An introduction to relational databases and other modern data technologies for statisticians can be found in Murrell [112].

R has two ways to access information in relational databases. There is a standard interface called ODBC, in the RODBC package [89], and there is an R protocol for writing database interfaces, in the DBI package [122]. In addition to these packages you need drivers for your specific database, and in most cases the database software itself. The survey package will only read from the database, it will never modify it. The data for the survey can be in a single database table, or in a 'view' that joins variables from multiple database tables. Creating a view that joins database tables is likely to be much faster than merging data sets in R.

The database-backed design objects behave the same way as other survey design objects, with just a few differences. The most obvious difference is that analyses of subsets of database-backed designs will occasionally report zero or NA results for empty categories of a factor rather than just omitting the category.

The survey data may already be stored in a database, or you may have to put it there. The example in this appendix is the multiple imputation data from NHANES III [153], discussed further in chapter 9. These data are available as plain text files with SAS code to read them. To create the database I exported the core file and five multiple imputation files from SAS, used read.xport() to read each file into R, then dbWriteTable() to save each file as a database table.

```
library(RSQLite)
sqlite <- dbDriver("SQLite")
conn <- dbConnect(sqlite, "imp.db")
imp <- read.xport("imp1.xpt")
dbWriteTable(conn, name="imp1", value=imp)
imp <- read.xport("imp2.xpt")
dbWriteTable(conn, name="imp2", value=imp)
```

I then used SQL commands

```
create view set1 as
  select * from core inner join imp1 using(SEQN)
```

to create five database views set1, set2, ..., set5 that merged the core file with each of the imputation files.

For the 2007 BRFSS data the size of the data set made it more difficult to create the database, and I used a server with several gigabytes of memory to read the data from the SAS XPORT format with read.xport() and write it to SQLite with dbWriteTable(). With 430,000 records, this data set is close to the limit for interactive analysis on a small computer: analyses will run, but even simple estimates take a minute or so. Since BRFSS is stratified by state at the first stage it is legitimate to work with a subset of the data when analyzing only one or a few states. For example, to analyze California's BRFSS data we can create a database view containing only the data for California (FIPS code 6), and supply this as the data argument to svydesign():

```
dbSendQuery(conn, "create view california as
 select * from brfss where X_STATE == 6")
ca_brfss <- svydesign(id=~X_PSU, weight=~X_FINALWT,
data="california",  dbtype="SQLite", dbname="brfss07.db")
```

It is also possible to create views to represent subsets of a data set. The SIPP data for a single wave of sampling come as a data set with one record per person for four months. The weights for a single month add up to the US population. Analyzing households rather than individuals requires using the subset of the data with just the "household reference person", that is, the subset where epppnum is equal to ehrefper.

```
library(RSQLite)
sqlite<-dbDriver("SQLite")
conn<-dbConnect(sqlite,"~/SIPP/sipp.db")
dbSendQuery(conn, "create view household as
 select * from wave1sub where epppnum==ehrefper")
dbDisconnect(conn)
```

The resulting database is on the web site for the book.

D.2 SETTING UP DATABASE INTERFACES

D.2.1 ODBC

On Windows systems ODBC drivers for Microsoft Access are built in, and ODBC drivers for other databases are usually included with the database software. The ODBC system can also run on Unix and Linux systems, and is included in recent versions of Apple's OS X.

The ODBC system uses names, called Data Source Names (DSNs), rather than file names, since a database could involve multiple files on more than one computer. An ODBC Adminstrator program (in the Control Panel on Windows and in the Utilities folder on OS X) is used to specify a DSN for the database. That DSN is then used to refer to the database inside R.

In the call to svydesign(), the data argument is a string, the name of the database table or view, the dbname argument is the DSN, and the dbtype argument is "ODBC".

```
library(RODBC)
odbc_nhanes <- svydesign(id=~SDPPSU6, strat=~SDPSTRA6,
    weight=~WTPFQX6, nest=TRUE, data="set1", dbname='nhanes3',
    dbtype="ODBC")
```

D.2.2 DBI

To use the R-DBI family of interfaces you need the DBI package and a package with the interface for your specific software. At the time of writing, interface packages are available for SQLite, Oracle, MySQL, PostgreSQL, and JDBC. At the time of writing, only the SQLite [66] and JDBC interfaces have been tested with the survey package.

In the call to svydesign(), the data argument is the name of the database table or view, the dbtype argument is the name of the database interface, dbname is the name by which the interface knows the specific database, and any other arguments needed to connect to the database can be supplied.

The simplest of these databases is SQLite [61], a small, fast, public-domain program that was designed to provide database facilities to other programs. It is not suitable for databases with many millions of rows, but will handle most surveys perfectly well. A SQLite database is stored in a single file, so the dbname argument is just the path to this file.

```
library(RSQLite)
sqlite_nhanes <- svydesign(id=~SDPPSU6, strat=~SDPSTRA6,
    weight=~WTPFQX6, nest=TRUE, data="set1",
    dbname='~/nhanes/imp.db', dbtype="SQLite")
```

MySQL [113] is a popular, larger scale, open-source, relational database system. A MySQL database is identified by the computer it is running on and a name similar to an ODBC DSN. For a MySQL database the command might look like

```
library(RMySQL)
mysql_nhanes <- svydesign(id=~SDPPSU6, strat=~SDPSTRA6,
    weight=~WTPFQX6, nest=TRUE, data="set1", dbname='nhanes3',
    dbtype="MySQL", user="thomas", host="abacus",
    password="bx2#de3Q")
```

to connect to a database called nhanes3 on a remote computer called abacus, with the specified user name and password. It is obviously undesirable to have the password exposed in an analysis script, and the RMySQL package has other options to store this information in configuration files.

Appendix E

EXTENDING THE PACKAGE

The range of regression models that might be of interest for complex survey data is almost unlimited, and only the most straightforward and frequently used ones have been implemented so far. The survey package provides some facilities for adding new regression models. As a case study, we will consider a negative binomial model.

E.1 A CASE STUDY: NEGATIVE BINOMIAL REGRESSION

An user of the survey package asked if there was a way to fit negative binomial regression models, to data where the outcome variable is a count. Negative binomial regression is not implemented in the survey package, but it is implemented for simple random samples in the MASS package [181].

A negative binomial regression model says that Y_i comes from a negative binomial distribution with mean given by

$$\log E[Y] = \log \mu_i = \alpha + X_i \beta \tag{E.1}$$

---

*

Complex Surveys: A Guide to Analysis Using R. By Thomas Lumley
Copyright © 2010 John Wiley & Sons, Inc.

and with variance
$$\text{var}[Y] = \mu_i + \mu_i^2/\theta. \tag{E.2}$$

The likelihood for one observation is

$$L(\mu_i, \theta) = \frac{\Gamma(Y_i + \theta)}{\Gamma(\theta)\Gamma(Y_i + 1)} \left(\frac{\theta}{\theta + \mu_i}\right)^{\theta} \left(\frac{\mu_i}{\mu_i + \theta}\right)^{Y_i}.$$

There are several options for this sort of count data:

1. Since the regression equation for the mean is the same as in a Poisson regression model, it would be possible to use svyglm() to fit a Poisson working model. Since the standard errors and p-values in svyglm() are design-based, they would be valid even for negative binomial data, which has more outliers than Poisson data. If the working model correctly described the mean, the regression coefficients would be consistent estimates of the same population quantity as for a negative binomial regression, otherwise they would be estimates of different, but still reasonable, population quantities.

2. For a replicate-weights design it is possible to use withReplicates() to estimate standard errors. It is then only necessary to compute correctly weighted point estimates. This strategy was used for the replicate-weights version of svyolr().

3. svymle() will fit sampling-weighted maximum likelihood estimates if given the likelihood and first derivative. This is more work than using replicate weights, but is likely to result in faster computation and can be used for designs where the survey package does not provide replicate weights (e.g., two-phase designs).

For some models a fourth strategy is to modify existing code that fits the model with simple random sampling so that the influence functions needed for a linearization estimator of standard error are returned and can be passed to svyrecvar(). This fourth strategy does not appear useful for the negative binomial model, but was used in writing the linearization version of svyolr().

A similar approach will work for many parametric models, in particular for parametric models for censored or truncated data such as the tobit model from econometrics. None of these strategies will work for all models, however. In particular, mixed models of any sort require estimates of covariance parameters and these need explicit weighting using the second-order sampling probabilities.

E.2   USING A POISSON MODEL

As an example we will examine data on lifetime number of sexual partners, collected by NHANES 2003–2004, and how it varies with age, gender, and race/ethnicity. The data are supplied by NCHS in a file sxq_c.xpt and can be read in and merged with the demographic data as in Section B.2.1. To compute the total number of sexual

```
demo <- read.xport("demo_c.xpt")
sxq <- read.xport("sxq_c.xpt")
merged <- merge(demo, sxq, by='SEQN')
merged$total <- with(merged, ifelse(RIAGENDR==2,
    SXQ100+SXQ130, SXQ170+SXQ200))
merged$total[merged$SXQ020==2] <- 0
merged$total[merged$total>2000] <- NA
merged$age <- merged$RIDAGEYR/25
des <- svydesign(id=~SDMVPSU,strat=~SDMVSTRA,weights=~WTINT2YR,
    nest=TRUE, data=merged)
```

Figure E.1   Loading and merging the example data from NHANES 2003–2004

partners when this is non-zero we need to add two variables, and these variables are different for men and women. We also need to set "refused" values to NA and assign a zero value. We also rescale the age variable so that the "zero" value for log(age) is at age 25 rather than age 2.7. The code is in Figure E.1.

The first possibility is to use design-based Poisson regression to fit the same log-linear model for the mean but with different weights. We can compare a model-based version of negative binomial regression using glm.nb() from the MASS package with the design-based Poisson regression using svyglm() (Figure E.2). If the point estimates are qualitatively similar there may be little need to go to the effort of implementing negative binomial regression. Even if they are different, Poisson regression may still be a useful summary of the associations in the data.

The glm.nb() function accepts a weights argument. Rather than passing the sampling weights WT2INT to the function, we scale the weights to sum to the sample size. This scaling often leads to better numerical stability when using existing functions with survey data, and also results in the model-based standard error estimates being at least of the right order of magnitude.

The coefficients from the unweighted negative binomial model, the weighted negative binomial model and the design-based Poisson regression are all broadly similar, as shown in Figure E.3. In this example, Poisson regression using svyglm() appears to be a good solution to the modelling problem.

E.3   REPLICATE WEIGHTS

The function glm.nb() in the MASS package fits negative binomial regression and accepts weights as an argument. Looking at the code shows that the weights are used to multiply individual contributions to the loglikelihood, so that the point estimates will be correct if the weights are frequency weights.

The first step is to convert the survey design object to a replicate-weights object if it is not already of that form, using as.svrepdesign(). We can now use withReplicates(), as shown in Figure E.4. The quoted expression is evaluated

```
library(MASS)
des <- update(des, scaledweights = WTINT2YR/mean(WTINT2YR))

model0 <- glm.nb(
    total~factor(RIAGENDR)*(log(age)+factor(RIDRETH1)),
    data=model.frame(des))
model1 <- glm.nb(
    total~factor(RIAGENDR)*(log(age)+factor(RIDRETH1)),
    data=model.frame(des), weights=scaledweights)
model2 <- svyglm(
    total~factor(RIAGENDR)*(log(age)+factor(RIDRETH1)),
    design=des, family=quasipoisson)
```

Figure E.2   Negative binomial regression (assuming independent sampling) compared with design-based Poisson regression

```
> round(cbind(coef(model0), coef(model1), coef(model2)), 2)
                                           [,1]   [,2]   [,3]
(Intercept)                                2.35   2.29   2.44
factor(RIAGENDR)2                         -0.91  -0.80  -0.95
log(age)                                   1.06   1.07   0.80
factor(RIDRETH1)2                          0.03   0.08   0.05
factor(RIDRETH1)3                          0.07   0.09   0.06
factor(RIDRETH1)4                          1.00   0.82   0.71
factor(RIDRETH1)5                         -0.01   0.06   0.01
factor(RIAGENDR)2:log(age)                -1.26  -1.22  -0.96
factor(RIAGENDR)2:factor(RIDRETH1)2        0.04  -0.18  -0.16
factor(RIAGENDR)2:factor(RIDRETH1)3        0.68   0.60   0.63
factor(RIAGENDR)2:factor(RIDRETH1)4       -0.03   0.06   0.18
factor(RIAGENDR)2:factor(RIDRETH1)5        0.64   0.38   0.41
```

Figure E.3   Estimates from model-based negative-binomial regression without weights and with weights, and from design-based Poisson regression

```
repldesign <- as.svrepdesign(des)
negbin <- withReplicates(repldesign,
   quote(coef(glm.nb(
     total~factor(RIAGENDR)*(log(age)+factor(RIDRETH1)),
     weights=.weights)))))
```

Figure E.4   Negative binomial regression using `withReplicates()`

with each set of replicates weights. It calls `glm.nb()` and returns the fitted coefficients. The point estimates obtained with the replicate-weights analysis are identical
to those from a weighted, model-based analysis but the design-based standard errors
are mostly larger, increasing by up to a factor of two.

   In a large sample the coefficients will not vary much between replicates, so time
may be saved by specifying starting values for the iterative procedure that fits the
model to each replicate. Figure E.5 illustrates this. The model is fitted with the
sampling weights, and the regression coefficients and θ estimated from that fit are
given as starting values. In this example the time saving was only about 10%, but
with larger data sets the savings can be greater.

```
model <- glm.nb(
    total~factor(RIAGENDR)*(log(age)+factor(RIDRETH1)),
    weights=weights(des), data=model.frame(des))
init.coef<-coef(model)
init.theta<-model$theta
negbin2 <- withReplicates(repldesign,
   quote(coef(glm.nb(
   total~factor(RIAGENDR)*(log(age)+factor(RIDRETH1)),
   weights=.weights, init.theta=init.theta, start=init.coef))))
```

Figure E.5   Supplying starting values to speed up `withReplicates()`

E.4   LINEARIZATION

The function `svymle()` will maximize a sampling-weighted estimate of a population
loglikelihood or other additive objective function. The user must supply the loglikelihood for a single observation and its first derivative with respect to its parameters.
It is then possible to give a model formula for any of these parameters.

   For the negative binomial model the loglikelihood is

$$\ell_i = \log \Gamma(Y_i + \theta) - \log \Gamma(\theta) - \log \Gamma(Y_i + 1)$$
$$+ \theta \log \theta + Y_i \log(\mu_i) - (\theta + Y_i) \log(\theta + \mu_i)$$

The derivative with respect to μ_i is

$$\frac{\partial \ell}{\partial \mu_i} = \frac{Y_i}{\mu_i} - \frac{\theta + Y_i}{\theta + \mu_i}.$$

The derivative with respect to θ involves the digamma function $\psi(\,)$, the derivative of the logarithm of the gamma function. This is available in R as `digamma()`.

$$\frac{\partial \ell}{\partial \theta} = \psi(\theta + Y_i) - \psi(\theta) + \log(\theta) + 1 - \log(\theta + \mu_i) - \frac{Y_i + \theta}{\mu_i + \theta}.$$

We will be specifying a linear model for $\eta = \log \mu$, so the derivative with respect to μ needs to be multiplied by $\partial \mu / \partial \eta = \mu$

The R code for the loglikelihood and its derivatives in Figure E.6 are straightforward translations of the equations, except that a trick is needed to ensure that $Y_i \log \mu_i$ evaluates as zero rather than undefined when $Y_i = 0$. The call to `svymle()` specifies the loglikelihood and derivative functions, and then specifies formulas for each parameter. There is only an intercept for θ, which is constant, but there is a full model formula for μ.

Starting values for iterative estimation are required with `svymle()`. Any valid values would be accepted, such as $\theta = 1$ and 0 for all the regression coefficients, but convergence will be more reliable with good starting values.

In this case `glm.nb()` would produce very good starting values. If it were not available, regression coefficients from a Poisson regression model would be reasonable starting values for the regression coefficients, and a starting value for θ could be estimated by substituting the mean of the fitted values and the variance of the residuals into equation E.2. Since θ cannot be negative and its value is not that far from zero (0.8, with a standard error of 0.3), reparametrizing in terms of $\log \theta$ might also give better estimation.

The coefficient estimates and standard errors from estimation with `svymle()` and `withReplicates()` are given in Figure E.7. The point estimates are identical (to more figures than given), and the standard errors are very similar. The standard errors are mostly about twice as large as from the model-based negative-binomial regression, so that ignoring the sampling design would cause seriously unreliable inference.

```
loglik <- function(y, theta, eta) {
    mu<-exp(eta)
    (lgamma(theta + y) - lgamma(theta) - lgamma(y + 1) +
      theta * log(theta) + y * log(mu + (y == 0))
      - (theta + y) * log(theta + mu))
    }

deta<- function(y, theta, eta) {
    mu <- exp(eta)
    dmu <-  y/mu - (theta+y)/(theta+mu)
    dmu*mu
}

dtheta <- function(y, theta, eta) {
    mu <- exp(eta)
    digamma(theta + y) - digamma(theta) + log(theta) + 1
      - log(theta + mu) - (y + theta)/(mu + theta)
  }

score<-function(y, theta, eta) {
    cbind(dtheta(y, theta,eta), deta(y, theta, eta))
}

modl1<-glm.nb(
    total~factor(RIAGENDR)*(log(age)+factor(RIDRETH1)),
    data=model.frame(des), weights=scaledweights)

nlmfit<-svymle(loglike=loglik, grad=score, design=des,
    formulas=list(theta=~1,
    eta=total~factor(RIAGENDR)*(log(age)+factor(RIDRETH1))),
    start=c(modl1$theta, coef(modl1)), na.action="na.omit")
```

Figure E.6   Fitting a negative binomial model with svymle()

```
> round(cbind(coef(nlmfit),SE(nlmfit), c(NA,coef(negbin)),
   c(NA, SE(negbin))),2)
                                             [,1]  [,2]   [,3] [,4]
theta.(Intercept)                            0.81  0.30    NA   NA
eta.(Intercept)                              2.29  0.16   2.29 0.16
eta.factor(RIAGENDR)2                       -0.80  0.18  -0.80 0.18
eta.log(age)                                 1.07  0.23   1.07 0.24
eta.factor(RIDRETH1)2                        0.08  0.15   0.08 0.15
eta.factor(RIDRETH1)3                        0.09  0.18   0.09 0.18
eta.factor(RIDRETH1)4                        0.82  0.30   0.82 0.30
eta.factor(RIDRETH1)5                        0.06  0.38   0.06 0.41
eta.factor(RIAGENDR)2:log(age)              -1.22  0.26  -1.22 0.27
eta.factor(RIAGENDR)2:factor(RIDRETH1)2     -0.18  0.26  -0.18 0.26
eta.factor(RIAGENDR)2:factor(RIDRETH1)3      0.60  0.19   0.60 0.20
eta.factor(RIAGENDR)2:factor(RIDRETH1)4      0.06  0.37   0.06 0.38
eta.factor(RIAGENDR)2:factor(RIDRETH1)5      0.38  0.44   0.38 0.46
```

Figure E.7   Results of negative binomial model using svymle() and withReplicates(). The estimates of θ were not extracted in withReplicates() and are given as NA.

References

1. J. A. Anderson. Separate sample logistic discrimination. *Biometrika*, 59:19–35, 1972.

2. Alusio J. D. Barros and Vânia Hirakata. Alternatives for logistic regression in cross-sectional studies: an empirical comparison of models that directly estimate the prevalence ratio. *BMC Medical Research Methodology*, 3(21), 2003.

3. Debrabata Basu. An essay on the logical foundations of survey sampling. In *Foundations of Statistical Inference*, pages 203–242. Holt, Rinehart & Winston, Toronto, 1971.

4. Richard A. Becker, William S. Cleveland, and Ming-Jen Shyu. The visual design and control of Trellis displays. *Journal of Computational and Graphical Statistics*, 5(123–155), 1996.

5. Richard A. Becker, Allan R. Wilks, and Ray Brownrigg. *mapdata: Extra Map Databases*, 2007. R package version 2.0-23.

6. Richard A. Becker, Allan R. Wilks, Ray Brownrigg, and Thomas P Minka. *maps: Draw Geographical Maps*, 2008. R package version 2.0-40.

7. David R. Bellhouse. Computing methods for variance estimation in complex surveys. *Journal of Official Statistics*, 1:323–329, 1985.

8. David A. Binder. On the variances of asymptotically normal estimators from complex surveys. *International Statistical Review*, 51:279–292, 1983.

9. David A. Binder. Fitting Cox's proportional hazards models from survey data. *Biometrika*, 79:139–147, 1992.

10. Yvonne M. M. Bishop, Stephen E. Fienberg, and Paul W. Holland. *Discrete Multivariate Analysis: Theory and Practice*. MIT Press, Cambridge, 1975.

11. Ørnulf Borgan, Bryan Langholz, Sven Ove Samuelsen, Larry Goldstein, and Janice Pogoda. Exposure stratified case-cohort designs. *Lifetime Data Analysis*, 6(1):39–58, 2000.

12. Norman E. Breslow. Statistics in epidemiology: the case-control study. *J. Amer. Stast. Assoc.*, 91(433):14–28, 1996.

13. Norman E. Breslow and Nilanjan Chatterjee. Design and analysis of two-phase studies with binary outcome applied to Wilms' tumor prognosis. *Applied Statistics*, 48:457–68, 1999.

14. Norman E. Breslow and Nicholas E. Day. *Statistical Methods in Cancer Research (Vol. 1): The Analysis of Case-control Studies*. World Health Organization [Distribution and Sales Service], 1980.

15. Norman E. Breslow and Nicholas E. Day. *Statistical Methods in Cancer Research: the design and analysis of cohort studies*, volume II. IARC Press, http://www.iarc.fr, 1987.

16. Norman E. Breslow, Thomas Lumley, Christie M. Ballantyne, Lloyd E. Chambless, and Michal Kulich. Improved Horvitz-Thompson estimation of model parameters from two-phase stratified samples: applications in epidemiology. *Statistics in Biosciences*, 1, 2009. forthcoming.

17. Norman E. Breslow, Thomas Lumley, Christie M Ballantyne, Lloyd E Chambless, and Michal Kulich. Using the whole cohort in the analysis of case-cohort data. *American Journal of Epidemiology*, 169(11):1398–1405, 2009.

18. Norman E. Breslow, Brad McNeney, and Jon A. Wellner. Large sample theory for semiparametric regression models with two-phase, outcome dependent sampling. *The Annals of Statistics*, 31(4):1110–1139, 2003.

19. Norman E. Breslow and Jon A. Wellner. Weighted likelihood for semiparametric models and two-phase stratified samples, with application to Cox regression. *Scandinavian Journal of Statistics*, 34:86–102, 2007.

20. Norman E. Breslow and Jon A. Wellner. A Z-theorem with estimated nuisance parameters and correction note for 'Weighted likelihood for semiparametric models and two-phase stratified samples, with application to Cox regression'. *Scand J. Statist*, 34:186–192, 2008.

21. Cynthia A. Brewer. *Designed Maps: A Sourcebook for GIS Users*. ESRI Press, 2008.

22. Kenneth R. W. Brewer and Muhammad Hanif. *Sampling with Unequal Probabilities*. Springer-Verlag, New York, 1983.

23. California Health Interview Survey. *CHIS 2005 Adult Public Use File, Release 1*. UCLA Center for Health Policy Research, Los Angeles, CA, January 2007.

24. California Health Interview Survey. *CHIS 2005 Methodology Series: Report 5 – Weighting and Variance Estimation*. UCLA Center for Health Policy Research, Los Angeles, CA, 2007.

25. California Health Interview Survey. *Methodology Brief: Weighting and Estimation of Variance in the CHIS Public Use Files*. UCLA Center for Health Policy Research, Los Angeles, CA, 2007.

26. Peter J. Cameron. Hadamard matrices. In *The Encylopedia of Design Theory*. 2005. `http://designtheory.org/library/encyc/`.

27. Angelo J Canty and Anthony C Davison. Resampling-based variance estimation for labour force surveys. *The Statistician*, 48:379–391, 1999.

28. D. B. Carr, R. J. Littlefield, W. L. Nicholson, and J. S. Littlefield. Scatterplot matrix techniques for large N. *Journal of the American Statistical Association*, 82:424–436, 1987.

29. Dan Carr, Nicholas Lewin-Koh, and Martin Maechler. *hexbin: Hexagonal Binning Routines*, 2008. R package version 1.14.0.

30. Nancy Cartwright. *Hunting Causes and Using Them: Approaches in Philosophy and Economics*. Cambridge University Press, June 2007.

31. Centers for Disease Control and Prevention (CDC). *Behavioral Risk Factor Surveillance System Operational and User's Guide*. U.S. Department of Health and Human Services, Centers for Disease Control and Prevention, Atlanta, GA, 2006. version 3.0.

32. John M. Chambers. *Software for Data Analysis: Programming with R*. Springer, New York, 2008.

33. John M. Chambers, William S. Cleveland, Beat Kleiner, and Paul A. Tukey. *Graphical Methods for Data Analysis*. Wadsworth, 1983.

34. Ronald Christensen. *Log-Linear Models and Logistic Regression*. Springer-Verlag, New York, 2nd edition, 1997.

35. William S. Cleveland. *Visualizing Data*. Hobart Press, Lafayette, IN, 1994.

36. Nancy R. Cook, Stephen R. Cole, and Charles H. Hennekens. Use of a marginal structural model to determine the effect of aspirin on cardiovascular mortality in the Physicians' Health Study. *Am. J. Epidemiol.*, 155(11):1045–1053, 2002.

37. Peter Dalgaard. *Introductory Statistics with R*. Springer, New York, 2nd edition, 2008.

38. GJ D'Angio, N Breslow, JB Beckwith, A Evans, H Baum, A de Lorimer, D Ferbach, E Hrabovsky, G Jones, P Kelalis, et al. Treatment of Wilms' tumor. Results of the Third National Wilms Tumor Study. *Cancer*, 64(2):349–60, 1989.

39. James A. Deddens and Martin R. Petersen. Re: "estimating the relative risk in cohort studies and clinical trials of common outcomes". *American Journal of Epidemiology*, 159:213–214, 2004. [letter].

40. Jean-Claude Deville and Carl-Erik Särndal. Calibration estimators in survey sampling. *Journal of the American Statistical Association*, 87:376–382, 1992.

41. Jean-Claude Deville, Carl-Erik Särndal, and Olivier Sautory. Generalized raking procedures in survey sampling. *Journal of the American Statistical Association*, 88:1013–1020, 1993.

42. A Dorfman and R Valliant. Quantile variance estimators in complex surveys. *Proceedings of the Section on Survey Research Methodology*, pages 866–871, 1993.

43. William H. DuMouchel and Greg J. Duncan. Using sample survey weights in multiple regression analyses of stratified samples. *Journal of the American Statistical Association*, 78(383):535–543, 1983.

44. J Durbin. A note on the application of Quenouille's method of bias reduction to the estimation of ratios. *Biometrika*, 46:477–80, 1959.

45. Paul Elliott, Jon Wakefield, Nicola Best, and David Briggs, editors. *Spatial Epidemiology: Methods and Applications.* Oxford Univ. Press, Oxford, 2001.

46. V. M. Estevao and C.-E. Särndal. Borrowing strength is not the best technique within a wide class of design-consistent estimators. *Journal of Official Statistics*, 20:645–660, 2004.

47. John Fox. *An R and S-Plus Companion to Applied Regression.* Sage Publications, Thousand Oaks, CA, USA, 2002.

48. Constantine E. Frangakis and Donald B. Rubin. Principal stratification in causal inference. *Biometrics*, 58(1):21–29, 2002.

49. Michael Friendly. A fourfold display for 2 by 2 by k tables. Technical Report 217, York University, Psychology Department, 1994.

50. Michael Friendly. *Visualizing Categorical Data.* SAS Institute, Cary, NC, 2000.

51. Fumio Funaoka, Hiroshi Saigo, Randy R. Sitter, and Tsutom Toida. Bernoulli bootstrap for stratified multistage sampling. *Survey Methodology*, 32(2):151–156, 2006.

52. Malay Ghosh and J. N. K. Rao. Small area estimation: An appraisal (Disc: P76-93). *Statistical Science*, 9:55–76, 1994.

53. Sander Greenland, James M. Robins, and Judea Pearl. Confounding and collapsibility in causal inference. *Statistical Science*, 14:29–46, 1999.

54. DM Gren, NE Breslow, JB Beckwith, FZ Finkelstein, PG Grundy, PRM THomas, T Kim, S Shochat, GM Haase, ML Ritchey, PP Kelalis, and GJ D'Angio. Comparison between single-dose and divided-dose administration of dactinomycin and doxorubicin for patients with Wilms' tumor: a report from the National Wilms' Tumor Study Group. *J Clin Oncol*, 16:237–245, 1998.

55. M. J. Gunter, D. R. Hoover, H. Yu, S Wassertheil-Smoller, T. E. Rohan, J. E. Manson, J. Li, G. Y. Ho, G. L. Anderson, R. C. Kaplan, T. G. Harris, B. V. Howard, J. Wylie-Rosett, R. D. Burk, and H. D. Strickler. Insulin, insulin-like growth factor-i, and risk of breast cancer in postmenopausal women. *J Natl Cancer Inst*, 101(1):48–60, 2009.

56. Jaroslav Hájek. Asymptotic theory of rejective sampling with varying probabilities from a finite population. *The Annals of Mathematical Statistics*, 35(4):1491–1523, 1964.

57. Jaroslav Hájek. Discussion of Basu (1971). In V. P. Godambe and D. A. Sprott, editors, *Foundations of Statistical Inference.* Holt, Rinehart, & Winston, Montreal, 1971.

58. Mark A. Harrower and Cynthia A. Brewer. Colorbrewer.org: An online tool for selecting color schemes for maps. *The Cartographic Journal*, 40(1):27–37, 2003.

59. H. O. Hartley and J. N. K. Rao. Sampling with unequal probabilities and without replacement. *The Annals of Mathematical Statistics*, 33:350–374, 1962.

60. Miguel Hernán, Babette Brumback, and James M. Robins. Marginal structural models to estimate the causal effect of zidovudine on the survival of HIV-positive men. *Epidemiology*, 11:561–570, 2000.

61. Hipp, Wyrick, & Company, Inc. *SQLite database*, 2007. version 3.4.1.

62. Paul W. Holland. Statistics and causal inference. *Journal of the American Statistical Association*, 81(396):945–960, 1986.

63. Daniel G. Horvitz and Donovan J. Thompson. A generalization of sampling without replacement from a finite universe. *J. Amer. Statist. Assoc.*, 47:663–685, 1952.

64. Peter J. Huber. The behavior of maximum likelihood estimates under nonstandard conditions. In *Proceedings of the Fifth Berkeley Symposium on Mathematical Statistics and Probability*, volume 1, pages 221–223, Berkeley, CA, 1967. University of California Press.

65. Cary T. Isaki and Wayne A. Fuller. Survey design under the regression superpopulation model. *Journal of the American Statistical Association*, 77(377):89–96, 1982.

66. David A. James. *RSQLite: SQLite interface for R*, 2008. R package version 0.6-9.

67. David R. Judkins. Fay's method for variance estimation. *Journal of Official Statistics*, 6:223–239, 1990.

68. David R. Judkins, David Morganstein, Paul Zador, Andrea Piesse, Brandon Barrett, and Pushpal Mukhopadhyay. Variable selection and raking in propensity scoring. *Statistics in Medicine*, 26:1022–1033, 2007.

69. G. Kalton and I. Flores-Cervantes. Weighting methods. *Journal of Official Statistics*, 19(2):81–97, 2003.

70. F. Kamangar, Y. L. Qiao, M. J. Blaser, X. D. Sun, J. Katki, J. H. Fan, G. I. Perez-Perez, C. C. Abnet, S. D. Mark, P. R. Taylor, and S. M. Dawsey. Helicobacter pylori and oesophageal and gastric cancers in a prospective study in China. *British Journal of Cancer*, 96(1):172–6, 2007.

71. Hormuzd A. Katki. *NestedCohort: Survival Analysis for Cohorts with Missing Covariate Information*, 2007. R package version 1.0-1.

72. A. Kaur, G. P. Patil, and C. Taillie. Ranked set sampling: an annotated bibliography. *Environmental and Ecological Statistics*, 2:25–54, 1995.

73. Scott Keeter, Carolyn Miller, Andrew Kohut, Robert M. Groves, and Stanley Presser. Consequences of reducing nonresponse in a national telephone survey. *Public Opinion Quarterly*, 64(2):125–148, 2000.

74. Jae Kwang Kim, J. Michael Brick, Wayne A. Fuller, and Graham Kalton. On the bias of the multiple-imputation variance estimator in survey sampling. *J. R. Statist Soc. B*, 68(3):509–521, 2006.

75. Leslie Kish. *Survey Sampling*. John Wiley & Sons, New York, 1965.

76. David G. Kleinbaum and Mitchel Klein. *Survival Analysis: a self-learning text*. Springer, New York, 2nd edition, 2005.

77. G. G. Koch, D. H. Freeman, and J. L. Freeman. Strategies in the multivariate analysis of data from complex surveys. *International Statistical Review*, 43:59–78, 1975.

78. G. G. Koch, C. M. Tangen, J. W. Jung, and I. A. Amara. Issues for covariance analysis of dichotomous and ordered categorical data from randomized clinical trials and non-parametric strategies for addressing them. *Statistics in Medicine*, 17:1863–92, 1998.

79. Roger Koenker. *Quantile Regression*. Cambridge University Press, Cambridge, 2005.

80. Roger Koenker. *quantreg: Quantile Regression*, 2008. R package version 4.23.

81. Edward L. Korn and Barry I. Graubard. Confidence intervals for proportions with small expected number of positive counts estimated from survey data,. *Survey Methodology*, 24:193–201, 1998.

82. Edward L. Korn and Barry I. Graubard. Scatterplots with survey data. *The American Statistician*, 52(1):58–69, 1998.

83. Edward L. Korn and Barry I. Graubard. *Analysis of Health Surveys*. Wiley, New York, 1999.

84. Phillip S. Kott. Using calibration weighting to adjust for non-response and coverage errors. *Survey Methodology*, 32:133–142, 2006.

85. Michal Kulich and D.Y. Lin. Improving the efficiency of relative-risk estimation in case-cohort studies. *Journal of the American Statistical Association*, 99(467):832–844, 2004.

86. Diego Kuonen. Saddlepoint approximations for distributions of quadratic forms in normal variables. *Biometrika*, 86(4):929–935, 1999.

87. Päivi Kurttio, Laina Salonen, Taina Ilus, Juka Pekkanen, Eero Pukkala, and Anssi Auvinen. Well water radioactivity and risk of cancers of the urinary organs. *Environmental Research*, 102(3):333–338, 2006.

88. Joseph H. Kutkuhn. *Estimating Absolute Age Composition of California Salmon Landings*. Number 120 in Fish Bulletin. Marine Resources Branch, California Department of Fish and Game, 1963.

89. Michael Lapsley and B. D. Ripley. *RODBC: ODBC Database Access*, 2008. R package version 1.2-3.

90. C. R. Lee, K. E. North, M. S. Bray, D. J. Couper, G. Heiss, and D. C. Zeldin. CYP2J2 and CYP2C8 polymorphisms and risk of cardiovascular events: the Atherosclerosis Risk in Communities (ARIC) study. *Pharmacogenet. Genomics*, 17(5):349–358, 2007.

91. C. R. Lee, K. E. North, M. S. Bray, D. J. Couper, G. Heiss, and D. C. Zeldin. Cyclooxygenase polymorphisms and risk of cardiovascular events: the Atherosclerosis Risk in Communities (ARIC) study. *Clin. Pharmacol. Ther.*, 83(1):52–60, 2008.

92. Risto Lehtonen and Ari Veijanen. Logistic generalized regression estimators. *Survey Methodology*, 24:51–55, 1998.

93. Nicholas J. Lewin-Koh, Roger Bivand, contributions by Edzer J. Pebesma, Eric Archer, Stéphane Dray, David Forrest, Patrick Giraudoux, Duncan Golicher, Virgilio Gómez Rubio, Patrick Hausmann, Thomas Jagger, Sebastian P. Luque, Don MacQueen, Andrew Niccolai, and Tom Short. *maptools: Tools for reading and handling spatial objects*, 2008. R package version 0.7-15.

94. Steff Lewis and Mike Clarke. Forest plots: trying to see the wood and the trees. *BMJ*, 322:1479–80, 2001.

95. D. Y. Lin. On fitting Cox's proportional hazards models to survey data. *Biometrika*, 87(1):37–47, 2000.

96. Danyu Lin and Zhiling Ying. Cox regression with incomplete covariate measurements. *Journal of the American Statistical Association*, 88:1341–1349, 1993.

97. Dennis V. Lindley. Basu's elephant. In *Encyclopedia of Statistical Sciences*. John Wiley & Sons, 2006.

98. Thomas Lumley. *rmeta: Meta-analysis*. R package version 2.14.

99. Thomas Lumley. Analysis of complex survey samples. *Journal of Statistical Software*, 9(1):1–19, 2004. R package verson 2.2.

100. Thomas Lumley. *mitools: Tools for multiple imputation of missing data*, 2008. R package version 2.0.

101. Thomas Lumley. *survey: Analysis of complex survey samples*, 2008. R package version 3.10-1.

102. Stefan Ma and Chit-Ming Wong. Estimation of prevalence proportion rates. *International Journal of Epidemiology*, 28:176, 1999. [letter].

103. Steven D. Mark and Hormuzd A. Katki. Specifying and implementing nonparametric and semiparametric survival estimators in two-stage (nested) cohort studies with missing case data. *Journal of the American Statistical Association*, 101(474):460–471, 2006.

104. P J McCarthy. Replication: an approach to the analysis of data from complex surveys. Series 2 14, National Center for Health Statistics, Hyattsville, MD, 1966.

105. Peter McCullagh. Regression models for ordinal data. *Journal of the Royal Statistical Society. Series B (Methodological)*, 42(2):109–142, 1980.

106. Doug McIlroy, Ray Brownrigg, and Thomas P Minka. *mapproj: Map Projections*, 2008. R package version 1.1-7.2.

107. G. A. McIntyre. A method for unbiased selective sampling, using ranked sets. *Australian Journal of Agricultural Research*, 3:385–390, 1952.

108. David Meyer, Achim Zeileis, and Kurt Hornik. *vcd: Visualizing Categorical Data. R package version 1.1-1.*, 2008.

109. Leyla Mohadjer and Lester R. Curtin. Balancing sample design goals for the National Health and Nutrition Examination Survey. *Survey Methodology*, 34:119–126, 2008.

110. Mark S. Monmonier. *Mapping It Out: Expository Cartography for the Humanities and Social Sciences.* University of Chicago Press, 1993.

111. Paul Murrell. *R Graphics.* Chapman & Hall/CRC, 2005.

112. Paul Murrell. *Introduction to Data Technologies.* Chapman & Hall, 2009.

113. MySQL AB. *MySQL database*, 2005. version 5.0.

114. V. Nambi, R. C. Hoogeveen, L. Chambless, Y. Hu, H. Bang, J. Coresh, H. Ni, E. Boerwinkle, T. Mosley, R. Sharrett, A. R. Folsom, and C. M. Ballantyne. Lipoprotein-associated phospholipase A2 and high-sensitivity C-reactive protein improve the stratification of ischemic stroke risk in the Atherosclerosis Risk in Communities (ARIC) study. *Stroke*, 2008. Dec 18 (online).

115. Bin Nan. Efficient estimation for case-cohort studies. *The Canadian Journal of Statistics / La Revue Canadienne de Statistique*, 32(4):403–419, 2004.

116. National Center for Health Statistics. *Plan and operation of the second National Health and Nutrition Examination Survey, 1976–1980.* Number 15 in Series 1, Programs and collection procedures. National Center for Health Statistics, 1981.

117. National Center for Health Statistics. Variance estimation for person data using SUDAAN and the National Health Interview Survey (NHIS) public use person data files, 1995–1994. Downloaded from http://www.nber.org/nhis/1994/docs/sudaan50.txt, 1995.

118. Marian L. Neuhouser, Lesley Tinker, Pamela A. Shaw, Dale Schoeller, Sheila A. Bingham, Linda Van Horn, Shirley A. A. Beresford, Bette Caan, Cynthia Thomson, Suzanne Satterfield, Lew Kuller, Gerardo Heiss, Ellen Smit, Gloria Sarto, Judith Ockene, Marcia L. Stefanick, Annlouise Assaf, Shirley Runswick, and Ross L. Prentice. Use of recovery biomarkers to calibrate nutrient consumption self-reports in the Women's Health Initiative. *Am. J. Epidemiol.*, 167(10):1247–1259, 2008.

119. Erich Neuwirth. *RColorBrewer: ColorBrewer palettes*, 2007. R package version 1.0-2.

120. Jerzy Neyman. On the two different aspects of the representative method: The method of stratified sampling and the method of purposive selection. *Journal of the Royal Statistical Society*, 97(4):558–625, 1934.

121. National Institute of Standards and Technology. *Counties and Equivalent Entities of the United States, Its Possessions, and Associated Areas*. Number 6–4 in FIPS. US Department of Commerce, 1990.

122. R Special Interest Group on Databases (R-SIG-DB). *DBI: R Database Interface*, 2007. R package version 0.2-4.

123. G. P. Patil. Ranked set sampling. In Abdel H El-Sharawi and Walter W. Piegorsch, editors, *Encyclopedia of Environmetrics*, volume 3, pages 1684–1690. John Wiley & Sons, Chichester, 2002.

124. Judea Pearl. *Causality : Models, Reasoning, and Inference*. Cambridge University Press, Cambridge, March 2000.

125. Edzer J. Pebesma and Roger S. Bivand. Classes and methods for spatial data in R. *R News*, 5(2):9–13, November 2005.

126. Andrew Pickles, Graham Dunn, and José Luis Vázquez-Barquero. Screening for stratification in two-phase ('two-stage') epidemiological surveys. *Statistical Methods in Medical Research*, 4:73–89, 1995.

127. Ross L. Prentice. A case-cohort design for epidemiologic cohort studies and disease prevention trials. *Biometrika*, 73:1–11, 1986.

128. Ross. L. Prentice and Ron Pyke. Logistic disease incidence models and case-control studies. *Biometrika*, 66:403–412, 1979.

129. J. N. K. Rao. *Small Area Estimation*. John Wiley and Sons, Hoboken, NJ, 2003.

130. J. N. K. Rao and A. J. Scott. On chi-squared tests for multiway contigency tables with proportions estimated from survey data. *Annals of Statistics*, 12:46–60, 1984.

131. J. N. K. Rao and C. F. J. Wu. Bootstrap inference for sample surveys. *Proceedings of the Section on Survey Research Methodology*, pages 106–112, 1984.

132. J. N. K. Rao and C. F. J. Wu. Resampling inference with complex survey data. *Journal of the American Statistical Association*, 83(231–241), 1988.

133. J. N. K. Rao, W. Yung, and M. A. Hidiroglou. Estimating equations for the analysis of survey data using poststratification information. *Sankhyā, Series A*, 64(2):364–378, 2002.

134. James M. Robins, Sander Greenland, and Fu-Chang Hu. Estimation of the causal effect of a time-varying exposure on the marginal mean of a repeated binary outcome. *Journal of the American Statistical Association*, 94:687–700, 1999.

135. James M. Robins, Miguel Hernán, and Babette Brumback. Marginal structural models and causal inference in epidemology. *Epidemiology*, 11:550–560, 2000.

136. James M. Robins and Andrea Rotnitzky. Comment on Bickel and Kwon: 'Inference for semiparametric models: some questions and an answer.'. *Statistica Sinica*, 11(4):920–936, 2001.

137. James M. Robins, Andrea Rotnitzky, and Lue Ping Zhao. Estimation of regression coefficients when some regressors are not always observed. *Journal of the American Statistical Association*, 89:846–866, 1994.

138. James M. Robins and Naisyin Wang. Inference for imputation estimators. *Biometrika*, 87(1):113–124, 2000.

139. Bernard Rosner. *Fundamentals of Biostatistics*. Duxbury Press, 5th edition, 1999.

140. Donald B. Rubin. Estimating causal effects of treatment in randomized and non-randomized studies. *Journal of Educational Psychology*, 66(688–701), 1974.

141. Donald B. Rubin. Inference and missing data. *Biometrika*, 63(3):581–592, 1976.

142. Donald B. Rubin. Bayesian inference for causal effects: the role of randomization. *Annals of Statistics*, 6:34–58, 1978.

143. Donald B Rubin. Multiple imputation in sample surveys — a phenomenological Bayesian approach to non-response. *Proceedings of the Survey Research Methods Section, American Statistical Association*, pages 20–34, 1978.

144. Donald B. Rubin. *Multiple Imputation for Nonresponse in Surveys*. Wiley, New York, 1987.

145. Donald B. Rubin. Multiple imputation after 18 years. *Journal of the American Statistical Association*, 91:473–90, 1996.

146. Sven Øve Samuelsen, Hallvard Ånestad, and Anders Skrondal. Stratified case-cohort analysis of general cohort sampling designs. *Scandinavian Journal of Statistics*, 34:103–119, 2007.

147. Deepayan Sarkar. *lattice: Lattice Graphics*, 2008. R package version 0.17-13.

148. Deepayan Sarkar. *Lattice: Multivariate Data Visualization with R*. UseR. Springer, New York, 2008.

149. Carl-Erik Särndal. The calibration approach in survey theory and practice. *Survey Methodology*, 33(2):99–119, 2007.

150. Carl-Erik Särndal and Bengt Swensson. A general view of estimation for two phases of selection with applications to two-phase sampling and nonresponse. *International Statistical Review / Revue Internationale de Statistique*, 55(3):279–294, 1987.

151. Carl-Erik Särndal, Bengt Swensson, and Jan Wretman. *Model Assisted Survey Sampling*. Springer, New York, 1992.

152. Carl-Erik Sarndal, Bengt Swensson, and Jan H. Wretman. The weighted residual technique for estimating the variance of the general regression estimator of the finite population total. *Biometrika*, 76(3):527–537, 1989.

153. J. L. Schafer. Analyzing the NHANES III multiply imputed data set: Methods and examples. Technical report, National Center for Health Statistics, Hyattsville, MD, 2001.

154. J. L. Schafer. Multiple imputation models and procedures for NHANES III. Technical report, National Center for Health Statistics, Hyattsville, MD, 2001.

155. K. A. Schulman, J. A. Berlin, W. Harless, et al. The effect of race and sex on physicians' recommendations for cardiac catheterization. *N Engl J Med*, 340:618–626, 1999.

156. Gary G. Schwartz, Dora Il'yasova, and Anastasia Ivanova. Urinary cadmium, impaired fasting glucose, and diabetes in the NHANES III. *Diabetes Care*, 26:468–470, 2003.

157. L. M. Schwartz, S. Woloshin, and H. G. Welch. Misunderstandings about the effects of race and sex on physicians' referrals for cardiac catheterization. *N Engl J Med*, 341:279–283, 1999.

158. Alastair Scott and Chris Wild. On the robustness of weighted methods for fitting models to case-control data. *Journal of the Royal Statistical Society, Series B: Statistical Methodology*, 64(2):207–219, 2002.

159. Alastair J. Scott and Chris J. Wild. Calculating efficient semiparametric estimators for a broad class of missing-data problems. In E. P. Liski, J. Isotalo, J. Niemelä, S. Puntanen, and G. P. H. Styan, editors, *Festschrift for Tarmo Pukkila on his 60th Birthday*, pages 301–314. Dept. of Mathematics, Statistics and Philosophy, Univ. of Tampere, 2006.

160. Steven G. Self and Ross L. Prentice. Asymptotic distribution theory and efficiency results for case-cohort studies. *The Annals of Statistics*, 16:64–81, 1988.

161. Babubhai V. Shah and Akhil K. Vaish. Confidence intervals for quantile estimation from complex survey data. *Proceedings of the Section on Survey Research Methodology*, 2006.

162. Jun Shao and Randy R. Sitter. Bootstrap for imputed survey data. *Journal of the American Statistical Association*, 91(435):1278–1288, 1996.

163. Jun Shao and Dongsheng Tu. *The Jackknife and Bootstrap*. Springer-Verlag Inc, New York, 1995.

164. Philip J. Smith. Is two-phase sampling really better for estimating age composition? *J Amer Statist Assoc*, 84:916–921, 1989.

165. P. G. Szilagyi, G. Fairbrother, M. R. Griffin, R. W. Hornung, S. Donauer, A. Morrow, M. Altaye, Y. Zhy, S. Ambrose, K. M. Edwards, K. A. Poehling, G. Lofthus, M. Holloway, L. Finelli, M. Iwane, M. A. Staat, and the New Vaccine Surveillance Network. Influenza vaccine effectiveness among children 6 to 59 months of age during 2 influenza seasons: a case-cohort study. *Arch. Pediatri. Adolesc. Med*, 162(10):943–51, 2008.

166. I. B. Tager, S. T. Weiss, B. Rosner, and F. E. Speizer. Effect of parental cigarette smoking on the pulmonary function of children. *American Journal of Epidemiology*, 110(1):15–26, 1979.

167. Catherine M. Tangen and G. G. Koch. Complementary nonparametric analysis of covariance for logistic regression in a randomized clinical trial setting. *J. Biopharm Stat.*, 9(1):45–66, 1999.

168. Terry M. Therneau and Hongzhe Li. Computing the Cox model for case cohort designs. *Lifetime Data Analysis*, 5:99–112, 1999.

169. D. Roland Thomas and J. N. K. Rao. Small-sample comparisons of level and power for simple goodness-of-fit statistics under cluster sampling. *Journal of the American Statistical Association*, 82(398):630–636, 1987.

170. Godfrey H. Thomson. A hierarchy without a general factor. *British Journal of Psychology*, 8:271–281, 1916.

171. L L Thurstone. The vectors of mind. *Psychological Review*, 41:1–32, 1934. available at http://psychclassics.yorku.ca/Thurstone/.

172. Yves Tillé. *Sampling Algorithms*. Springer Series in Statistics. Springer, New York, 2006.

173. Yves Tillé and Alina Matei. *sampling: Survey Sampling*, 2008. R package version 2.0.

174. Anastasios A. Tsiatis, Marie Davidian, Min Zhang, and Xiaomin Lu. Covariate adjustment for two-sample treatment comparisons in randomized clinical trials: a principled yet flexible approach. *Statistics in Medicine*, 27(23):4658–77, 2008.

175. A J Tuyns and G Massé. Cancer of the oesophagus in Brittany: an incidence study in Ille-et-Vilaine. *Int J Epidemiol.*, 4(1):55–9, 1975.

176. A J Tuyns, G Péquignot, and O M Jensen. Le cancer de l'oesophage en Ille-et-Vilaine en fonction des niveaux de consommation d'alcool et de tabac. Des risques qui se multiplient. *Bulletin du Cancer*, 64:45–60, 1977.

177. Antony Unwin, Martin Theus, and Heike Hoffman. *Graphics of Large Datasets:Visualizing a million.* Springer, New York, 2006.

178. Richard Valliant. Poststratification and conditional variance estimation. *Journal of the American Statistical Association*, 88(421):89–96, 1993.

179. Aad W. van der Vaart. *Asymptotic Statistics.* Cambridge University Press, Cambridge, 2000.

180. Robert J. Vanderbei. On graphical representations of voting results. Technical report, Princeton University, 2008. `http://www.princeton.edu/~rvdb/papers_online_webpage.html`.

181. William N. Venables and Brian D. Ripley. *Modern Applied Statistics with S.* Springer, New York, fourth edition, 2002.

182. Matt Wand and Brian Ripley. *KernSmooth: Functions for kernel smoothing for Wand and Jones (1995)*, 2008. R package version 2.22-22.

183. Matt P Wand and M C Jones. *Kernel Smoothing.* Chapman & Hall, London, 1995.

184. Naisyin Wang and James M. Robins. Large-sample theory for parametric multiple imputation procedures. *Biometrika*, 85:935–948, 1998.

185. Halbert White. A heteroskedasticity-consistent covariance matrix and a direct test for heteroskedasticity. *Econometrica*, 48:817–838, 1980.

186. Hadley Wickham. *ggplot2: An implementation of the Grammar of Graphics*, 2008. R package version 0.7.

187. Leland Wilkinson. *The Grammar of Graphics.* Springer, New York, 2nd edition, 2005.

188. Kirk M. Wolter. *Introduction to Variance Estimation.* Springer-Verlag Inc, 2nd edition, 2007.

189. Ralph S. Woodruff. Confidence intervals for medians and other position measures. *J Amer Statist Assoc*, 57:622–627, 1952.

190. Changbao Wu and Randy R. Sitter. A model-calibration approach to using complete auxiliary information from survey data. *Journal of the American Statistical Association*, 96:185–193, 2001.

191. Scott L. Zeger and Kung-Yee Liang. Longitudinal data analysis for discrete and continuous outcomes. *Biometrics*, 42(1):121–130, 1986.

192. Min Zhang, Anastasios A. Tsiatis, and Marie Davidian. Improving efficiency of inferences in randomized clinical trials using auxiliary covariates. *Biometrics*, 64(3):707–15, 2008.

Author Index

Topic Index

R functions

WILEY SERIES IN SURVEY METHODOLOGY
Established in Part by WALTER A. SHEWHART AND SAMUEL S. WILKS

Editors: *Mick P. Couper, Graham Kalton, J. N. K. Rao, Norbert Schwarz, Christopher Skinner*
Editor Emeritus: *Robert M. Groves*

The *Wiley Series in Survey Methodology* covers topics of current research and practical interests in survey methodology and sampling. While the emphasis is on application, theoretical discussion is encouraged when it supports a broader understanding of the subject matter.

The authors are leading academics and researchers in survey methodology and sampling. The readership includes professionals in, and students of, the fields of applied statistics, biostatistics, public policy, and government and corporate enterprises.

ALWIN · Margins of Error: A Study of Reliability in Survey Measurement
BETHLEHEM · Applied Survey Methods: A Statistical Perspective
*BIEMER, GROVES, LYBERG, MATHIOWETZ, and SUDMAN · Measurement Errors in Surveys
BIEMER and LYBERG · Introduction to Survey Quality
BRADBURN, SUDMAN, and WANSINK ·Asking Questions: The Definitive Guide to Questionnaire Design—For Market Research, Political Polls, and Social Health Questionnaires, *Revised Edition*
BRAVERMAN and SLATER · Advances in Survey Research: New Directions for Evaluation, No. 70
CHAMBERS and SKINNER (editors · Analysis of Survey Data
COCHRAN · Sampling Techniques, *Third Edition*
CONRAD and SCHOBER · Envisioning the Survey Interview of the Future
COUPER, BAKER, BETHLEHEM, CLARK, MARTIN, NICHOLLS, and O'REILLY (editors) · Computer Assisted Survey Information Collection
COX, BINDER, CHINNAPPA, CHRISTIANSON, COLLEDGE, and KOTT (editors) · Business Survey Methods
*DEMING · Sample Design in Business Research
DILLMAN · Mail and Internet Surveys: The Tailored Design Method
FULLER · Sampling Statistics
GROVES and COUPER · Nonresponse in Household Interview Surveys
GROVES · Survey Errors and Survey Costs
GROVES, DILLMAN, ELTINGE, and LITTLE · Survey Nonresponse
GROVES, BIEMER, LYBERG, MASSEY, NICHOLLS, and WAKSBERG · Telephone Survey Methodology
GROVES, FOWLER, COUPER, LEPKOWSKI, SINGER, and TOURANGEAU · Survey Methodology, *Second Edition*
*HANSEN, HURWITZ, and MADOW · Sample Survey Methods and Theory, Volume 1: Methods and Applications
*HANSEN, HURWITZ, and MADOW · Sample Survey Methods and Theory, Volume II: Theory
HARKNESS, VAN DE VIJVER, and MOHLER · Cross-Cultural Survey Methods
KALTON and HEERINGA · Leslie Kish Selected Papers
KISH · Statistical Design for Research
*KISH · Survey Sampling
KORN and GRAUBARD · Analysis of Health Surveys
LEPKOWSKI, TUCKER, BRICK, DE LEEUW, JAPEC, LAVRAKAS, LINK, and SANGSTER (editors) · Advances in Telephone Survey Methodology

*Now available in a lower priced paperback edition in the Wiley Classics Library.

LESSLER and KALSBEEK · Nonsampling Error in Surveys

LEVY and LEMESHOW · Sampling of Populations: Methods and Applications, *Fourth Edition*

LUMLEY · Complex Surveys: A Guide to Analysis Using R

LYBERG, BIEMER, COLLINS, de LEEUW, DIPPO, SCHWARZ, TREWIN (editors) · Survey Measurement and Process Quality

MAYNARD, HOUTKOOP-STEENSTRA, SCHAEFFER, VAN DER ZOUWEN · Standardization and Tacit Knowledge: Interaction and Practice in the Survey Interview

PORTER (editor) · Overcoming Survey Research Problems: New Directions for Institutional Research, No. 121

PRESSER, ROTHGEB, COUPER, LESSLER, MARTIN, MARTIN, and SINGER (editors) · Methods for Testing and Evaluating Survey Questionnaires

RAO · Small Area Estimation

REA and PARKER · Designing and Conducting Survey Research: A Comprehensive Guide, *Third Edition*

SARIS and GALLHOFER · Design, Evaluation, and Analysis of Questionnaires for Survey Research

SÄRNDAL and LUNDSTRÖM · Estimation in Surveys with Nonresponse

SCHWARZ and SUDMAN (editors) · Answering Questions: Methodology for Determining Cognitive and Communicative Processes in Survey Research

SIRKEN, HERRMANN, SCHECHTER, SCHWARZ, TANUR, and TOURANGEAU (editors) · Cognition and Survey Research

SUDMAN, BRADBURN, and SCHWARZ · Thinking about Answers: The Application of Cognitive Processes to Survey Methodology

UMBACH (editor) · Survey Research Emerging Issues: New Directions for Institutional Research No. 127

VALLIANT, DORFMAN, and ROYALL · Finite Population Sampling and Inference: A Prediction Approach